PROCESS TECHNOLOGIES FOR WATER TREATMENT

Earlier Brown Boveri Symposia

PROCESS TECHNOLOGIES FOR WATER TREATMENT

Edited by

Samuel Stucki
Asea Brown Boveri, Ltd.
Baden, Switzerland

PLENUM PRESS • NEW YORK AND LONDON

Library of Congress Cataloging in Publication Data

Brown Boveri Symposium on Process Technologies for Water Treatment (1987: Brown Boveri Research Center)
   Process technologies for water treatment / edited by Samuel Stucki.
      p.   cm.
   "Proceedings of the tenth Brown Boveri Symposium on Process Technologies for Water Treatment, held September 21–22, 1987, in Baden, Switzerland" — T.p. verso.
   Includes bibliographical references and inex.
     ISBN 978-1-4684-8558-5   ISBN 978-1-4684-8556-1 (eBook)
     DOI 10.1007/978-1-4684-8556-1

   1. Water — Purification — Congresses. I. Stucki, Samuel. II. Title.
TD433.B76 1987   88-21918
628.1′62 — dc19   CIP

Proceedings of the Tenth Brown Boveri Symposium on Process Technologies for Water Treatment, held September 21–22, 1987, in Baden, Switzerland

© 1988 Plenum Press, New York
Softcover reprint of the hardcover 1st edition 1988

A Division of Plenum Publishing Corporation
233 Spring Street, New York, N.Y. 10013

Foreword

The Brown Boveri Scientific Symposia by now are part of a firmly established tradition. This is the tenth event in a series which was initiated shortly after Corporate Research was created as a separate entity in our company; the symposia are held every other year. The themes have been:

1969   Flow Research on Blading
1971   Real-Time Control of Electric Power Systems
1973   High-Temperature Materials in Gas Turbines
1975   Nonemissive Electrooptic Displays
1977   Current Interruption in High-Voltage Networks
1979   Surges in High-Voltage Networks
1981   Semiconductor Devices for Power Conditioning
1983   Corrosion in Power Generating Equipment
1985   Computer Systems for Process Control
1987   Process Technologies for Water Treatment

The tenth event in an uninterrupted series that by now goes back almost 20 years is a good opportunity to make a few remarks on the guiding rules that have governed our symposia.

Why have we chosen these titles? At the outset we established certain selection criteria; we felt that a subject for a symposium should fulfill the following three requirements:

- It should characterize a part of an established discipline; in other words, it should describe an area of scholarly study and research.

- It should be of current interest in the sense that important results have recently been obtained and considerable research is still being undertaken in the world's scientific community.

- It should bear some relation to the scientific and technological activity of the company.

Let us look at the requirement "current interest": Some of the topics on our list of titles have been the object of research for several decades, some even from the beginning of the century. One might wonder, then, why such fields could be called particularly timely in the 70's and 80's. Let me make a few remarks on this subject. The reason is the following:

Experience shows that scientific progress in most areas does not occur at a constant rate — it comes in waves that are often followed by periods where successful research activity comes to a standstill. The waves are often sparked by an external event that may come from quite an unexpected corner. Our symposium subjects have always been chosen so as to coincide with such a wave, and the present one is no exception, as I will point out.

Along with electricity, *water* is one of the basic commodities which will be needed in ever increasing quantities as the industrialization and urbanization of our world progresses. The challenge of rising demand can only be met if we learn to apply all suitable scientific and technical resources available to us in stepping up the productivity of process technologies for turning low grade water into pure water.

Moreover, at the fast moving frontiers where new industrial products are developed and manufactured, we observe that whenever the complexity and sophistication of those products reach a higher level, so does the demand on the purity of the processing water required for making them. Manufacture of semiconductor integrated circuits provides a clear example. The removal of *all* foreign matter from the processing water is the ultimate goal. What now limits further progress is not so much our ability to improve water purification, but rather the sensitivity of the available measuring instruments for the reliable detection of residual matter.

But our theme does not end with ultrapure water. There is a global awareness now of the dangers of water contamination from industrial and other waste. The serious consequences that this may have for humanity call for drastic remedial measures. Therefore, it is essential to develop more efficient and economic processes to remove toxic or hazardous substances from effluents. Failure to do so may endanger highly developed industries as the added cost of waste water treatment may make production uneconomic. Today, improvements in handling waste are becoming as important as product development in maintaining competitiveness.

As in previous meetings, the number of participants was limited in order to maintain the character of a specialist meeting of restricted size, and it was with much regret that we were forced to disappoint many who wished to participate but who could not be accommodated. We hope that the publication of the present volume is a partial consolation for those whom we did not have the pleasure of welcoming as our guests.

The Symposium was attended by 110 participants from 14 countries. It was both an honour and a pleasure to welcome these scientists and engineers from so many different parts of the world. Their willingness to travel to Baden to spend two full days with us was a challenge as well as an obligation to us as organizers, and we sincerely hope that the expectations which prompted them to attend were fulfilled.

To conclude, we should like to take this opportunity of expressing our sincere gratitude to every Symposium participant. We hope they consider the time spent with us to have been worthwhile. Thanks are due primarily to the authors for having spared no effort in preparing their papers: the contents of this volume reflect the high quality of their work. We thank also the participants in the discussions, both formal and informal, and the editor of these proceedings.

The selection of the theme, the layout of the program and the contacting of the speakers were the responsibility of Dr. S. Stucki, head of the Electrochemistry and Membrane Separations Group at our Research Center. His careful and competent preparation was instrumental in the success of the Symposium. Our thanks also go to Mr. H. Wilhelm and his staff for the smooth running of the administrative side of the meeting.

On January 1, 1988, the Brown Boveri Company merged with ASEA AB, Sweden, to become Asea Brown Boveri, or ABB, a multinational corporation. In this volume, the affiliation of the speakers and participants has been listed as at the time of the symposium.

A.P. Speiser
Scientific Adviser of the Executive Board
of Asea Brown Boveri
formerly Director of Corporate Research
of Brown Boveri

Preface

The goal of the 10th Brown Boveri Symposium was to give a summary of the state of the art of advanced physical-chemical water treatment technologies, their principles and their applications in municipal and industrial water supplies. This volume contains the proceedings of the symposium which was held at the Brown Boveri Research Center on September 21 and 22, 1987.

Water treatment has become of increasing importance in a world of limited resources. The quality standards required for health in the case of drinking water, or for reliability in the case of industrial process water, necessitate improved technologies for the treatment of the water before use, as well as before discharge.

The main body of the symposium consists of 12 invited papers covering a broad range of subjects. The edited versions of the discussion sessions which were held after the oral presentation of the papers at the symposium are included after each chapter.

It was originally planned to divide this volume into two parts: the first dealing with the principles of various treatment techniques, followed by a second part looking at their applications. It turned out, however, that the distinction between "fundamental" and "applied" papers was very arbitrary, and therefore this grouping was abandoned. The applications of the various water treatment processes, such as membrane separation, UV, ozone or electrochemical treatment or their combinations, range from heavily polluted waste water streams from chemical plants to ultra-purification of water for microelectronics manufacturing processes.

The book opens with a key note lecture on water treatment by M.C. Kavanaugh, which emphasizes (referring to the problem of haloform removal as a very illustrative example) the importance of an improved chemical engineering approach to water treatment in guaranteeing the quality standards imposed by the health authorities.

Two chapters focus on membrane separation processes. R. Rautenbach and I. Janisch show that reverse osmosis (RO) membranes are capable of doing more than just removing ions. Removal of organic contaminations by commercial RO modules seems feasible as a new application of RO in the field of waste water treatment. In the following chapter an exhaustive review of all the possible electrodialytic processes is given by the inventor of electrodialysis, W. McRae. Electrodialysis, as the oldest membrane process for desalination of water, plays an important role in brackish water desalination and is specially suited for applications in remote or less developed areas of the world as it requires few chemicals.

Electrochemical water treatment has been used in the removing of heavy metals from the waste water produced in galvanic and related processes. G. Kreysa discusses the various types of electrochemical reactors which have been considered or used industrially for this purpose. With proper design of the electrochemical cells, electrochemical removal of heavy metals from waste water has proved to be a competitive process.

The state of the art of ozone generation by gas discharge is reviewed by U. Kogelschatz. The implementation of the well-understood physics of the gas discharge processes in technical equipment has led to a new generation of ozonizers with improved efficiency and, at the same time, enhanced space-time yield. These improvements are a prerequisite for the large-scale use of ozone in water technology.

A good knowledge of the chemistry of ozone in aqueous media is of practical importance in two ways: it gives the engineer the necessary data about ozone demand and ozonation time; and it provides a clue to the important problem of reaction products which have to be expected from reactions of ozone with pollutants. J. Hoigné gives an authoritative review of the complex aqueous chemistry of ozone.

H.P. Klein reviews the various uses of ozone in water treatment processes, emphasizing the roles of process technology as well as of ozone production technology for optimum performance of the process. If properly designed, ozonation alone or in combination with other processes can perform a number of purification stages which are necessary in reaching a given product quality from a given source.

Ultraviolet radiation is an alternative to chlorination or ozonation for disinfection purposes. The photochemical mechanisms of inactivation of microorganisms are discussed in the chapter written by C. v. Sonntag. UV radiation of 254 nm wavelength is found to be such an effective disinfectant because it selectively induces irreversible photochemical reactions within the vital DNA molecules of the microorganisms.

Water of extreme purity is required in various industrial processes to act as a chemical, solvent or as a rinsing agent. Krapf and Preisser show the complexity of the problems that arise if one is to guarantee the extremely high water purity standards required for microelectronic manufacturing. If the tolerable impurity levels are in the range of parts per billion (ppb) to parts per trillion (ppt), control of contamination becomes an intricate task comprising optimum design of the purification equipment, trace analysis and monitoring of all possible classes of contaminants, and flow design of the whole piping system.

A new process for the removal of trace organics from ultrapure water is presented in the chapter by S. Stucki and H. Baumann. The combination of in-situ produced ozone with UV radiation and mixed-bed polishing has proved to be a promising process for simultaneously controlling TOC and biological contamination in ultrapure water systems.

Production processes in the chemical industry and in related industries (e.g. pulp and paper) involve waste water that normally needs special treatment before it can be discharged into the environment or into a municipal sewage system. E. Plattner and C. Comninellis compare in their article the various possible approaches to removing soluble organic material from waste water. These involve either separation processes (adsorption, membrane separation) or oxidative transformation into a harmless end-product.

Power plants need huge quantities of water of diverse quality depending on how it is put to use (ranging from boiler feed make-up to flue-gas scrubbing solutions). The paper of L. Pelloni, A. Kyas and I. Reimer shows that, by proper nesting of the various water streams taking into account the range of purity levels present in a power station, considerable cost savings for the raw material "water" are possible.

The collection of articles in this volume exemplifies the challenge that water treatment poses. Water is a cheap commodity (quoted treatment costs range from a few cts/m^3 to more than 20 SFr /m^3) with, at the same time, very high purity standards (standard drinking water has a purity of 99.5%!). Process technologies for water purification involve delicate separation problems, i.e. the removal of small amounts of impurities from large volumes of water. However, the resources allocated to these processes are limited, which means that, from the multitude of possible technical solutions, the engineer has to solve the sometimes very delicate problem of choosing the most economic process or combination of processes. The chapters in this book clearly demonstrate the progress that has been made both in the physical chemistry and in the chemical engineering of water treatment processes. But they also leave a number of questions unanswered, which means that there is still need for further research and development in order to refine and improve water treatment processes.

I would like to express my thanks to all those who contributed to the success of the symposium and to the editing of the proceedings. First I am grateful to the authors of the papers for agreeing to do the work of preparing their oral presentations as well as writing up the articles reproduced here. I hope they will find that the result of their work, i.e. the present book, has been worth their extra efforts. In preparing the scientific program of the symposium, I was advised by Dr. C. Schüler. I greatly appreciated his encouraging enthusiasm. For their help before and during the symposium, I would like to thank Ms. B. Wullimann and all my colleagues of the Electrochemistry and Membrane Separations Group.

The editing of the book would have been impossible without the assistance of Ms. J. Nehring, who not only typed all the manuscripts, but also masterfully checked that all the references, captions, etc., were eventually brought into a coherent and correct form. Dr. S. Dingwall did an excellent job polishing our English and pointing out any linguistic inconsistencies in the articles and discussions. The camera-ready layout, including the sizing of the figures and the text, was prepared very skillfully by Mr. A. Miquel. Many thanks to all of them!

Baden, April 1988   S. Stucki

Contents

Contents

Participants

Mr. Jürg Aeppli
Wasserversorgung Zürich
Hardhof 9, Postfach
CH - 8023 Zürich
Switzerland

Dr. M. Bodmer
BBC Brown, Boveri & CO., Ltd.
FB TEZC.KC
CH - 5401 Baden
Switzerland

Mr. Leif Andersson
ASEA Stal AB
S - 61220 Finspång
Sweden

Mr. A. Bonnin
SRTI
283, Rue de la Minière
F - 78530 Buc
France

Mr. Kurt Bäckström
AB Electro-Invest
P.O.Box 3284
S - 10365 Stockholm
Sweden

Mr. Colin Bowler
Technical Manager
Ames Crosta Babcock Ltd.
Gregge Street
Heywood, Lancs. OL10 2DX
Great Britain

Dr. H. Baumann
Brown Boveri Research Center
Dept. CRBH.E
CH - 5405 Baden
Switzerland

Dr. Arthur Bradley
Fisons Plc.
London Road
Homes Chapel
Crewe, Cheshire
Great Britain

Dr. W. Blaser
Kernkraftwerk Leibstadt AG
Abteilung Überwachung
CH - 4353 Leibstadt
Switzerland

Dipl.-Ing. F.C. J. Brandt
Kernkraftwerk Gösgen-Däniken
Postfach
CH - 4658 Däniken
Switzerland

Mr. H. Brunner
BBC Brown, Boveri & Co., Ltd.
Dept. ISU
Postfach 8242
CH - 8050 Zürich
Switzerland

Prof. Dr. Ivan Dobrevsky
University of Chemical Technology
8010 Bourgas
Bulgaria

Ing. HTL Gérard Büschi
F. Hoffmann-La Roche & Co. AG
Bau 74, 2. OG-W
Postfach
Grenzacherstrasse 124
CH - 4002 Basel
Switzerland

Mr. P. Dyer-Smith
BBC Brown, Boveri & Co., Ltd.
Dept. ISU
Postfach 8242
CH - 8050 Zürich
Switzerland

Prof. Dr. Maurice Campagna
Direktor am Institut für
Festkörperforschung der
Kernforschungsanlage   Jülich GmbH
Postfach 1913
D - 5170 Jülich 1
Germany

Dr. M. Fischer
BBC Brown, Boveri & Co., Ltd.
Dept. ISW-RH
Postfach 8242
CH - 8050 Zürich
Switzerland

Dr. C. Comninellis
EPFL
Département de Génie Chimique
CH - 1024 Ecublens
Switzerland

Mr. Peter Francis
The Electricity Council
Research Centre
Capenhurst
Chester CH1 6ES
Great Britain

Dr. B. Czeska
Linnhoff GmbH - SRTI
Hauptstrasse 25
D - 7813 Staufen
Germany

Dr. T.P. Gasser
BBC Brown, Boveri & Co., Ltd.
KL
CH - 5401 Baden
Switzerland

Dipl.-Ing. A. Deuschle
Projektabteilung
GOEMA Dr. Götzelmann KG
Postfach 9 95
D - 7000 Stuttgart 1
Germany

Dr. G.H. Gessinger
BBC Brown, Boveri & Co., Ltd.
FB TEZ
CH - 5401 Baden
Switzerland

Mr. J.M. Giannone
Griffin Technics, Inc.
P.O. Box 330
178 Route 46
Lodi, NJ 07644
U.S.A.

Dr. Ir. J. Gons
Storck Wafilin
P.O. Box 13
NL - 8400 AA Gorredijk
The Netherlands

Dr. Rainer Graef
ROBERT BOSCH
Postfach 50
D - 7000 Stuttgart 1
Germany

Mr. M.W. Hannah
British Brown Boveri Ltd.
Darby House
Lawn Central
Telford
Shropshire TF3 4 JB
Great Britain

Mrs. Ann-Charlotte Harrysson
Fläkt Industri AB
Air Pollution Control
S - 35187 Växjö
Sweden

Mr. Dipl.-Ing. G. Hartmann
c/o German Water Engineering
GmbH
P.O. Box 1560
D - 4450 Lingen
Germany

Dr. K. Hermann
Chemische Rundschau
Dornackerstr. 39
CH - 4501 Solothurn
Switzerland

Dr. M. Hirth
Brown Boveri Research Center
Dept. CRBP.F
CH - 5405 Baden
Switzerland

Dr. Dieter Hody
Sulzer-Escher Wyss AG
Postfach
CH - 8023 Zürich
Switzerland

Prof. Dr. J. Hoigné
EAWAG
Überlandstrasse 133
CH - 8600 Dübendorf
Switzerland

Mrs. Carol Ruth James
Senior Engineer
6565 Oakwood Drive
Oakland, CA 94611
U.S.A.

Dipl.-Ing. I. Janisch
RWHT Aachen
Institut für Verfahrenstechnik
D - 5100 Aachen
Germany

Dr. H. Jodeit
Brown Boveri Research Center
Dept. CRBP.F
CH - 5405 Baden
Switzerland

Prof. O. Kedem
Weizmann Inst. of Science
Dept. of Membrane Research
P.O. Box 138
Rehovot 76101
Israel

Dr. T. Kaiser
BBC Brown, Boveri & Co., Ltd.
FB TEZC
CH - 5401 Baden
Switzerland

Dr. H.P. Klein
BBC Brown, Boveri & Co., Ltd.
Dept. ISW
Postfach 8242
CH - 8050 Zürich
Switzerland

Dipl.-Ing. (FH) Reinhard Kalbfuss
Thyssen Wassertechnik GmbH
Oelhafenstrasse 4-6
D - 6800 Mannheim 1
Germany

Dr. U. Kogelschatz
Brown Boveri Research Center
Dept. CRBP
CH - 5405 Baden
Switzerland

Mr. K. Kallenberg
Direktor
ROPUR AG
Postfach
Kirschgartenstrasse 12
CH - 4010 Basel
Switzerland

Dr. R. Kötz
Brown Boveri Research Center
Dept. CRBH.E
CH - 5405 Baden
Switzerland

Dr. Ali Karimi
Water Quality Specialist
Los Angeles Department
of Water and Power
P.O. Box 111, Room 1350
Los Angeles, CA 90051
U.S.A.

Mr. E. Krapf
IBM Werk Sindelfingen
Postfach 266
D - 7032 Sindelfingen
Germany

Dr. Michael C. Kavanaugh
James M. Montgomery
Consulting Engr., Inc.
501 Lennon Lane
Walnut Creek, CA 946 11
U.S.A.

Mr. Krebser
MMD Maschinenfabrik Meyer AG
Oeschbachstrasse 428
CH - 4707 Deitingen
Switzerland

Prof. Dr. G. Kreysa
Vice Director
Dechema-Institut
Theodor-Heuss-Allee 25
Postfach 97 01 46
D - 6000 Frankfurt/M. 97
Germany

Dr. A. Kyas
BBC Brown, Boveri & Co., Ltd.
FB TEZC.TV
CH - 5401 Baden
Switzerland

Mr. M. Kyburz
MMD Maschinenfabrik Meyer AG
Oeschbachstrasse 428
CH - 4707 Deitingen
Switzerland

Mr. Gaspar Lesznik
Griffin Technics, Inc.
P.O. Box 330
178 Route 46
Lodi, NJ 07644
U.S.A.

Mr. P.A. Liechti
BBC Brown, Boveri & Co., Ltd.
Dept. ISU-N
Postfach 8242
CH - 8050 Zürich
Switzerland

Dr. William G. Light
Director R & D
UOP Fluid Systems
10054 Old Grove Road
San Diego, CA 92131
U.S.A.

Mr. W. Lorch
Lorch Foundation
Buckwood
Fulmer Bucks SL3 6JN
Great Britain

Mr. Wayne A. McRae
P.O. Box 192
CH - 8053 Zürich
Switzerland

Dr. R.W. Meier
Brown Boveri Research Center
Dept. CRB
CH - 5405 Baden
Switzerland

Dr. A. Menth
BBC Brown, Boveri & Co., Ltd.
UB-VE
CH - 5401 Baden
Switzerland

Dr. E. Merz
BBC Brown, Boveri & Co., Ltd.
Dept. ISU
Postfach 8242
CH - 8050 Zürich
Switzerland

Dr. Kurt Meuli
Christ AG
Hauptstrasse 192
CH - 4147 Aesch
Switzerland

Mr. A. Miquel
Brown Boveri Research Center
Dept. CRBH.E
CH - 5405 Baden
Switzerland

Dr. G.R. Peyton
ENR Illinois Department
of Energy and Natural Resources
Aquatic Chemistry Section
2204 Griffith Drive
Champaign, IL 61820
U.S.A.

Prof. M. Mirbach
BBC Brown, Boveri & Co., Ltd.
FB TEZC.T
CH - 5401 Baden
Switzerland

Dr. H. Pfenninger
CIBA-GEIGY AG, Basel
K-674.712
CH - 4002 Basel
Switzerland

Mr. P. Müller
DRM
Dr. Müller AG
Alte Landstrasse 415
CH - 8708 Männedorf
Switzerland

Prof. Dr. E. Plattner
Département de Génie Chimique
EPFL
CH - 1024 Ecublens
Switzerland

Dr. F. Münzel
Im Weideli
CH - 8624 Grüt bei Wetzikon
Switzerland

Dr. Gerhard Pohl
Leiter der Anwendungstechnik
Degussa AG
Industrie- und Feinchemikalien
Postfach 1345
D - 6450 Hanau 1
Germany

Dr. L.J. Overman
N.V. KEMA
Utrechtsweg 310
NL - 6800 Arnhem
The Netherlands

Mr. R. Preisser
IBM Werk Sindelfingen
Postfach 266
D - 7032 Sindelfingen
Germany

Mr. L. Pelloni
BBC Brown, Boveri & Co., Ltd.
Dept. T-VN
CH - 5401 Baden
Switzerland

Mr. P.W. Prendiville
Senior Vice President
Camp Dresser & McKee, Inc.
One Center Plaza
Boston, MA 02108
U.S.A

Prof. Dr. R. Rautenbach
RWTH Aachen
Institut für Verfahrenstechnik
Turmstrasse 46
D - 5100 Aachen
Germany

DI Jorge A. Redondo
Dow Chemical Company
Filmtec Corporation
Industriestrasse 1
D - 7587 Rheinmünster 2
Germany

Prof. David W.T. Rippin
Technisch-Chemisches Laboratorium
ETH-Zentrum
CH - 8092 Zürich
Switzerland

Dr. G. Riva
Ozono Elettronica Inter. SpA.
Via Cesare Battisti 11
I - 20122 Milano
Italy

Mr. C.M. Robson
Camp Dresser & McKee
10101 Linn Station Road
Louisville, KY 40223
U.S.A.

Dr. B. Romacker
BBC Brown, Boveri & Co., Ltd.
KL
CH - 5401 Baden
Switzerland

Dr. A. Ruf
Institut für Verfahrens- und
Kältetechnik
ETH Zentrum
CH - 8092 Zürich
Switzerland

Mr. T. Samuel
Kernkraftwerk Beznau der NOK
KKB
Schlossweg 6
CH - 5312 Döttingen
Switzerland

Mr. E. Schenker
Eidgenössisches Institut
für Reaktorforschung
EIR-SIN
CH - 5303 Würenlingen
Switzerland

Mr. Werner Schildknecht
BBC Brown, Boveri & Co., Ltd.
Dept. VI
Mellingerstrasse 6
CH - 5401 Baden
Switzerland

Dr. W. Schlachter
BBC Brown, Boveri & Co., Ltd.
UB KWT
CH - 5401 Baden
Switzerland

Prof. Klaus Schneider
ENKA AG
Forschungsinstitut
Postfach
D - 8753 Obernburg
Germany

Dr. G. Schock
Brown Boveri Research Center
Dept. CRBH.E
CH - 5405 Baden
Switzerland

Dr. J. Stankovic
Energo Projekt Oour
Kneza Milosa 28
YU - 11000 Belgrade
Yugoslawia

Mr. Tom Schofield
Assistant Divisional Manager
Severn-Trent Water
Tame Division
156 Newhall Street
Birmingham B3 1SE
Great Britain

Dr. K. Stemmer
SANDOZ AG
Bau 301/153
Postfach
CH - 4002 Basel
Switzerland

Dr. C. Schüler
Brown Boveri Research Center
Dept. CRBH
CH - 5405 Baden
Switzerland

Prof. E.L. Stover
Oklahoma State University
Civil Engineering Dept.
P.O. Box 2056
Stillwater, OK 74078
U.S.A.

Prof. K.H. Simmrock
Universität Dortmund
Technische Chemie A
Vogelpothsweg
D - 4600 Dortmund 50
Germany

Dr. H. Strathmann
Fraunhofer Institut für
Grenzflächen-
und Bioverfahrenstechnik (IGB)
Nobelstrasse 12
D - 7000 Stuttgart 80
Germany

Mr. R. Simonet
MMD Maschinenfabrik Meyer AG
Oeschbachstrasse 428
CH - 4707 Deitingen
Switzerland

Dr. S. Stucki
Brown Boveri Research Center
Dept. CRBH.E
CH - 5405 Baden
Switzerland

Prof. Dr. Ambros P. Speiser
Brown Boveri Research Center
Dept. CRB
CH - 5405 Baden
Switzerland

Dr. R. Svoboda
BBC Brown, Boveri & Co., Ltd.
FB TEZC.K
CH - 5401 Baden
Switzerland

Mr. Mats Toll
Managing Director
AB Electro-Invest
P.O. Box 3284
S - 10365 Stockholm
Sweden

Mr. Kazuo Tomiie
Chlorine Engineers Corp., Ltd.
Research & Development Dept.
Shosen Mitsui Bldg. 3F
No. 1-1, Toranomon 2-Chome
Minato-Ku, Tokyo, 105
Japan

Mr. K. Toyoda
Chlorine Engineers Corp., Ltd.
Research & Development Dept.
Shosen Mitsui Bldg. 3F
No. 1-1, Toranomon 2-Chome
Minato-Ku, Tokyo, 105
Japan

Mr. Ir. E.A. van Naerssen
Head Dept. of Waterquality
Management
N.V. Regionaal Energiebedrijf
Dordrecht
Noordendijk 250
NL - 331 RR Dordrecht
The Netherlands

Mr. L. Vogel
BBC Brown, Boveri & Co., Ltd.
Dept. ISU
Postfach 8242
CH - 8050 Zürich
Switzerland

Prof. Dr. C. von Sonntag
Max-Planck-Institut für
Strahlenchemie
Stiftstrasse 34-36
D - 4330 Mühlheim/Ruhr 1
Germany

Dr. Voss
BBC Brown, Boveri & Cie. AG
Dept. GK/DV
Postfach 3 51
D - 6800 Mannheim 1
Germany

Dr. H.P. Walliser
CIBA–GEIGY AG
K-693.1.22
CH - 4002 Basel
Switzerland

Prof. Bao-Zhen Wang
Water Pullotion Control Res. Centre
Harbin Architectural and Civil
Engineering Institute
144 Dazhi Street
Harbin
China

Mr. Börje Wegemar
ASEA Atom
S - 72178 Västerås
Sweden

Chem. Ing. K. Wieck–Hansen
Power Plant
Nordkraft
P.O. Box 238
DK- 9200 Aalborg
Denmark

Dr. N. Wiegart
Brown Boveri Research Center
Dept. CRBP.P
CH - 5405 Baden
Switzerland

Water Treatment

Michael C. Kavanaugh

James M. Montgomery Consulting Engr., Walnut Creek, USA

1. Introduction

Water treatment in its broadest context can be defined as the subjection of water to an agent or process with the objective of transforming the source quality to meet application-specific criteria or standards. Water treatment has been a concern of all human settlements since earliest times. Prior to the industrial revolution, water treatment evolved based on empirical observations without a scientifically based understanding of the underlying mechanisms of treatment processes. In the public sector, the major objective of water treatment up to the 20th century, was providing a water of desirable aesthetic quality while at the same time protecting consumers from water born diseases. In the industrial sector, during this time, the primary concern was obtaining water in sufficient quantity and quality to permit expansion of the growing industrial centers.

In comparison to the predominance of empirical developments, the post-19th century era of water treatment has seen a rapid application of the principles of chemistry, materials science, microbiology and engineering science towards improvements in process efficiencies, optimization of existing processes, and the development of new processes for removal of contaminants from water. In the industrialized nations, these applications have led to a near eradication of acute health hazards caused by water born diseases. At the same time, advances in analytical chemistry and toxicological science have raised new concerns regarding potential chronic health hazards from drinking water containing trace (less than 0.1 mg/l) levels of organic chemicals, not of natural origin.

In the private sector during this period, demands have increased for higher quality water, increased efficiency in water use, minimization of the quantities of waste produced, and increased reliability both in water treatment and waste water treatment. These demands have both an economic incentive and an ethical imperative.

One of the dominant features of this transition from empirical methods to the use of a more scientifically based approach has been the growing application of the principles of process analysis to water treatment. A recent textbook on the subject[1] prepared by this author and his colleagues was structured to demonstrate the application of these principles to water treatment design. In spite of the improved understanding of the processes used in water treatment, however, many empirical rules still dominate the selection of design criteria, particularly in public sector water treatment applications.

Various factors have inhibited wider application of scientific and technical advances towards the optimization of design in water treatment applications. Usually a water treatment problem involves a multi-component system which has in the past rendered the use of mathematical models impractical. Each source of water has unique characteristics which can have a significant impact on the operations of specific processes. The situation is further exacerbated by the lack of source quality control. Water treatment processes must be designed to treat waters with a range of influent water quality normally not encountered in the chemical process industries.

Finally, there is the issue of cost. Water is by any measure an inexpensive commodity, when not in scarce supply. Treated water costs typically $ 0.5 - $1.00/m^3, which is equivalent to $ 0.0005/kg to $ 0.001/kg, a small cost in spite of large usage in comparison to such liquid products as gasoline ($ 1.00/kg) or bottled water ($ 1.00/kg). Given this exceptionally low price, the usual economic incentives for improving process efficiency are often lacking.

Nonetheless, the need for continuing advances in water treatment technology is strong, based on world-wide trends in both degradation of water quality and the growing scarcity of high quality water sources. In the United States, surface water in recent years has improved in some areas and become degraded in others[2]. With continued population growth, non point sources are becoming a growing cause of surface water contamination. In industrialized nations, which often rely extensively on ground water as a water supply source, contamination of this resource with synthetic organic chemicals is a growing problem[3].

In this context of increased concern over potential or real chronic health effects resulting from drinking water contaminated with trace levels of synthetic organic chemicals, the trend amongst regulatory agencies in the United States and elsewhere is to increase the number of parameters which must be measured and controlled. Table 1 summarizes the growth in the number of parameters regulated by the Environmental Protection Agency (EPA) in the United States over the past 10 years. As can be seen, EPA currently regulates 49 constituents, and expects to be regulating over a hundred constituents after 1990. In addition, because over 53,000 chemicals are currently in commercial production[9], more compounds may ultimately be added to the list.

Another impetus for improvements in water treatment technology is the growing scarcity of water in most arid regions of the world[4]. In these areas, water treatment facilities are pressed to provide adequate water quality from sources experiencing ever-increasing water quality degradation/contamination. Several areas of the U.S. are contemplating direct reuse of treated waste water despite public opposition. Any decreases in source water quality place increasing demands on the efficiency and reliability of water treatment technology.

Table 1:   Regulatory development in the United States: number of regulated constituents.

Year	Number of Constituents Regulated by USEPA
1914	2
1925	11
1942	19
1946	20
1962	30
1980	42
1987	49
1990	>100

In order to meet these increasing demands on water treatment processes in both the private and public sectors, creative application of process engineering principles is essential. In this paper, as the keynote address in this symposium on Process Technologies for Water Treatment, I propose to review briefly those process engineering principles which provide a framework for efficient advances in water treatment technologies. These principles will be illustrated by a review of process developments in technologies for control of synthetic organic chemicals in water.

2. Processing Engineering in Water Treatment

From a historical perspective, water treatment process technology in the 20th Century has advanced considerably in comparison to earlier empirical developments. As summarized by Baker[5], the 16th and 17th centuries saw the application of coagulation, gravity sedimentation, boiling, use of screens, and the use of sand filters for removal of particulate matter. Such processes were operated in a batch mode. Process development depended entirely on empirical evidence. Use of chemical agents increased during the 18th century with use of lime softening and chemical precipitation. Also, the adsorptive capabilities of powdered charcoal were discovered and utilized for removal of natural organic matter and taste and odor producing compounds. In the 19th century, the first large-scale engineered systems were introduced, primarily filtration systems utilized in production of potable water for municipalities. This era also saw the advent of continuous coagulation/sedimentation systems, and aeration for removal of volatile taste and odor-causing compounds. Additional advances included flotation and the use of biological processes for waste water treatment.

The 20th century evolution of water treatment technology has been marked by the discovery of new process technologies and the optimization of earlier technologies. Typical examples include: disinfection with chlorine, ozone and other oxidants; the use of ion exchange for removal of specific anions and cations; the discovery and application of membrane processes for removal of dissolved constituents; oxidation process technologies; continuous adsorption systems and the use of air stripping processes for removal of volatile compounds and ammonia in water and waste water treatment facilities. An overview of this historical development is given in Table 2.

From one perspective, 20th century developments in water treatment technology can be viewed as the culmination of translating natural phenomena into engineered systems, whereby the normally slow rates of reaction occurring in natural environments are accelerated in physical structures. Examples of this analogy, listed in Table 3, include coagulation of clays and other particulate matter in estuaries following the mixing of turbid river waters in estuaries, flow through porous media, leading to the use of sand and other coarse media for removal of particulate materials in engineered systems. Other examples include evaporation, which is mirrored by distillation processes, soil ion exchange processes, the precursor of inorganic and organic ion exchange materials, atmospheric UV oxidation, incorporated in photolytic oxidation processes in water, and animal membranes which were the precursors of synthetic membranes used in reverse osmosis or ultrafiltration processes.

Currently, water treatment process technology is used in a wide variety of applications, both in the public and private sectors. Some examples of the principal applications are shown in Table 4. Municipal potable water treatment must meet stringent water quality requirements usually imposed by public health agencies or other regulatory bodies. Ground water remediation through extraction and treatment is a widely accepted alternative in cleaning aquifiers contaminated by releases from hazardous waste sites. The major objective is the removal of organic contaminants which appear to be the predominant constituents currently causing ground water contamination in the U.S.[3]. Major private sector water treatment applications include production of industrial process water and treatment of process cooling

Table 2:   Historical overview of water treatment process development (after Baker[5]).

Period	Process Development
Pre-16th Century	• Coagulation, sedimentation, filtration in batch mode
	• Boiling
17th Century	• Distillation
	• Screening
	• Sand filters
18th Century	• Lime softening
	• Chemical precipitation
	• Infiltration
	• Adsorption on powdered charcoal
19th Century	• Large-scale filtration systems
	• Coagulation/sedimentation
	• Aeration
	• Flotation
	• Biological processes
20th Century	• Disinfection
	• Ion exchange
	• Membrane processes
	• Oxidation
	• Adsorption
	• Air stripping

Table 3:   Evolution of water treatment technologies based on analogies with natural phenomena.

Natural Phenomena	Engineered System
Coagulation of clays in estuaries	Coagulation/sedimentation
Ground water flow through porous media	Filtration
Evaporation	Distillation
Soil ion exchange	Inorganic ion exchangers
Atmospheric UV-oxidation	UV-oxidation in water
Foam formation	Flotation
Oxygen depletion in water	Biological processes
Animal membranes	Reverse osmosis/ultrafiltration

water. Both of these applications are dominated by increasingly stringent quality requirements.

Water treatment of municipal or industrial waste water for discharge to receiving waters or to sewers is a major water treatment process challenge. Municipal waste water treatment traditionally has relied on biological processes for oxidation of organic matter. Although physical/chemical treatment has been utilized in circumstances where a large fraction of the waste water is of industrial origin, biological processes are more widely used because they have been shown to be more reliable and cost effective for this application. Industrial waste water treatment is usually dominated by physical/chemical treatment processes because of the diversity of constituents that must be treated. More recent applications include water treatment for the production of non-potable water from waste water. In this application, an overriding issue is the ability of the processes to achieve a consistent water quality when treating a highly variable influent water quality[6]. Finally, a future water treatment application is the production of potable water in space stations. In this case, cost becomes a less important issue in comparison to reliability under potentially extreme conditions.

Each of these applications poses unique technical challenges to the water treatment engineer who can utilize an array of increasingly sophisticated tools to determine the most appropriate treatment processes for any given application and to improve the performance of existing processes to achieve savings in costs and an improved finished water quality.

3. Strategy of Process Design

In the broadest sense, process analysis consists of a sequence of decision points leading to the selection, design, construction and operation of a water treatment system[7]. Fig. 1 presents a logic diagram which describes the overall design sequence. The first issue a process engineer must address is definition of the problem. This includes characterizing the influent water quality, determining the quantity of water to be treated, defining treatment goals with respect to an average concentration of contaminants to be removed, and with respect to an acceptable level of reliability[6]. This stage might also include a definition of operational requirements with respect to the level of operational skills needed by the plant staff.

The second step in this sequence is the selection of alternatives for analysis. Specific

Table 4:   Range of water treatment applications.

Application	Dominant Feature
Municipal potable water treatment	Quality regulated, public sector primarily
Ground water remediation	Organic contaminants, small systems
Industrial process water treatment	Stringent quality requirements
Process cooling water	Economics of recycle
Waste water for discharge	Biological processes, sludge a problem
Waste water for non-potable reuse	Quality variability
Industrial waste water treatment	Wide diversity of constituents, high concentrations
Space stations	Reliability

technologies or subsystems must be identified, based on previous experience or based on knowledge of the physical and chemical properties of the contaminants and their impact on treatment options. At this stage, the potentially feasible technologies are identified and the relationships between these technologies specified.

Under normal circumstances, more than one alternative process technology may achieve the desired treatment goals. Thus, prior to an evaluation of alternatives, a preliminary screening stage is desirable. This could include preparation of steady state materials and energy balances, or use of simplified kinetic or equilibrium models to estimate potential removal efficiencies for the processes. Order of magnitude cost estimates might also be employed as well as engineering judgement to determine which alternatives should be subjected to more detailed analysis. This is the stage at which it is also determined whether or not pilot studies are appropriate for the purposes of verifying the expected performance of the technologies to be evaluated.

If the processes are well known, the project could be completed in a short time frame. At this point it is possible to prepare performance specifications and proceed directly to the purchase of the appropriate equipment from manufacturers. Construction, start up and operation of the treatment process then follow. This particular sequence is frequently used in industrial settings as well as in ground water remediation applications where the need for rapid response may be critical.

Assuming that bench and pilot studies are not needed, the subsequent steps in the design decision process would include selection of an appropriate model for describing the technologies. Numerous model types are available to the engineer, depending upon the nature of the process[8]. These include transport models based on transport phenomena, population balance models primarily used for particulate separation processes (e.g. filtration, flotation) and empirical models.

Following the selection of the model type, the level of detail for a specific model must be determined. Again, depending upon the amount of information available and the level of detail commensurate with the potential cost savings, a number of options are available to the engineer[8]. The most detailed approach is the use of multiple gradient models in which all expected phenomena are accounted for in the appropriate differential equations including concentration changes in three dimensions. This approach is primarily a research tool and is rarely used in process analysis because of the cost and the inherent inaccuracies of these transport models which neglect such phenomena as dispersion, and generally address concentration changes in only one dimension. This level of detail is widely used in reactor analysis and also the analysis of continuous processes.

Finally, the lowest level of detail is employed through use of macroscopic models in which scaling effects are neglected and concentrations are averaged over the entire volume of the process equipment. A combination of maximum gradient transport models with empirical models has been successfully used in the chemical process industry. Applications to water treatment process technology have been more limited due to the inhibiting factors discussed earlier.

Following selection of the appropriate analytical model, optimization studies can be completed. These would include sensitivity analyses for a range of values of process parameters and possible use of linear programming techniques to define the global or site specific optimum combination of processes and the optimum design criteria[7]. This optimization step can then be followed by a quantitative definition of acceptable overdesign factors (i.e. safety factors) which would account for the uncertainties in both the design process, in the values of the estimated process parameters and in the influent water quality.

In practice, this overdesign factor is usually selected empirically, based on previous experience. In the public sector, this often leads to excessively conservative designs. Upon completion of the optimization and overdesign analysis, specific design criteria would be selected followed by detailed mechanical, electrical, and structural design. The final steps include purchase of appropriate equipment and/or construction of the water treatment process, start-up and operations.

The process analysis component of the overall design and construction process, which would include decision points in the logic diagram (Fig. 1) through selection of the specific design criteria, typically represents approximately 5 to 20 percent of the total cost of the project. But its impact on total project costs is large, and usually represents a sound investment. In some industrial water treatment process design problems, the cost of this complete analysis may be excessive and consequently the equipment purchase route may be preferred. Whether or not investments should be made in the complete process analysis, which may include bench or pilot studies, is an important decision to be made at the preliminary screening stage.

4. Alternatives Analysis - Process Selection

Water treatment engineers must typically select processes which are capable of removing a wide range of constituents from the source water. In the alternatives selection process, engineers can rely on previous experience regarding removal of the constituents of concern. Alternatively, a process engineer can proceed to select the individual technologies which should be evaluated and included in treatment system alternatives based on the physical chemical properties of the constituents of concern.

In some cases it is convenient to utilize schematic selection diagrams which are based on empirical or theoretical evaluation of process performance. Fig. 2 illustrates the probable range

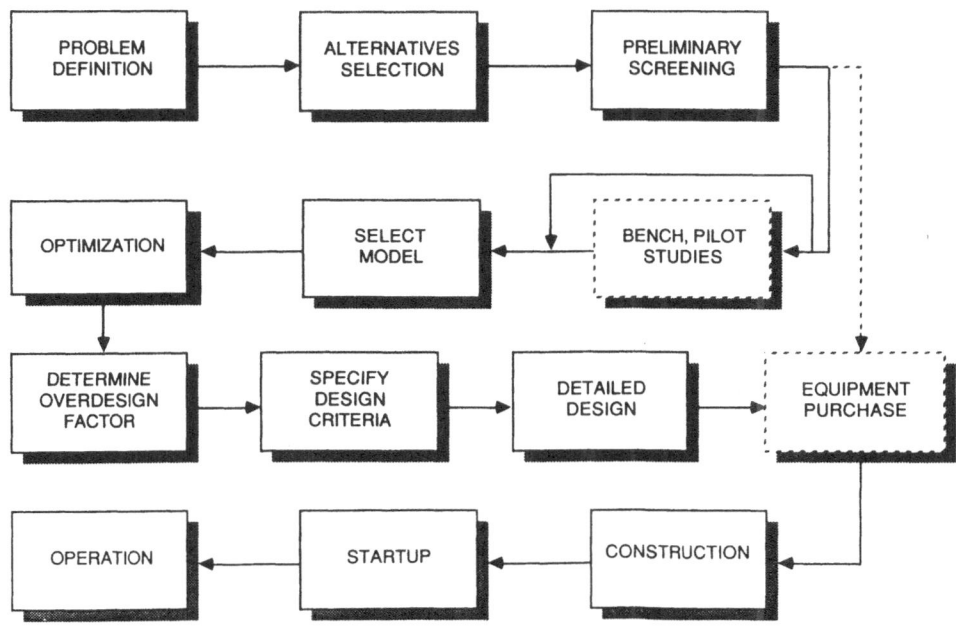

Fig. 1:   Logic diagram of design process in water treatment applications.

of generic separation processes modified on the basis of a familiar plot prepared by Dorr Oliver & Company. Shown are the primary factors affecting separation processes as a function of the size of the constituent of concern. For example Henry's constant allows an assessment of the suitability of air stripping for the removal of dissolved constituents, either volatile organic chemicals or dissolved gases. The size of a contaminant also determines whether or not microscreening might be a desirable process as opposed to gravity separation. The size and density are, in this case, the factors which determine the appropriate technologies for further evaluation.

Fig. 2 is only marginally useful, however, because separation processes generally involve complex interactions between process parameters and the contaminants to be removed. An example of a more detailed analysis for process selection in the case of particulate removal is shown in Fig. 3[10]. In this case, the combined effects of the size, the mass concentration, and number of particles have been evaluated to define the approximate regions in which specific technologies would be appropriate for particulate removal. It is assumed in this analysis that particulate matter is appropriately destabilized to permit effective coagulation or filtration. The boundaries separating the different regions of operation are based in part on empirical observations and theoretical considerations regarding the kinetics of the processes. Recent work in this area has further refined the applicable zones of process selection[11].

Similarly, a process selection diagram can be prepared for determining the appropriate aeration technologies which should be evaluated for the removal of organics in water. Such a diagram is shown in the Fig. 4 in which the desired percent removal and the volatility of the compound as characterized by the Henry's constant are shown[13]. Thus, depending upon the

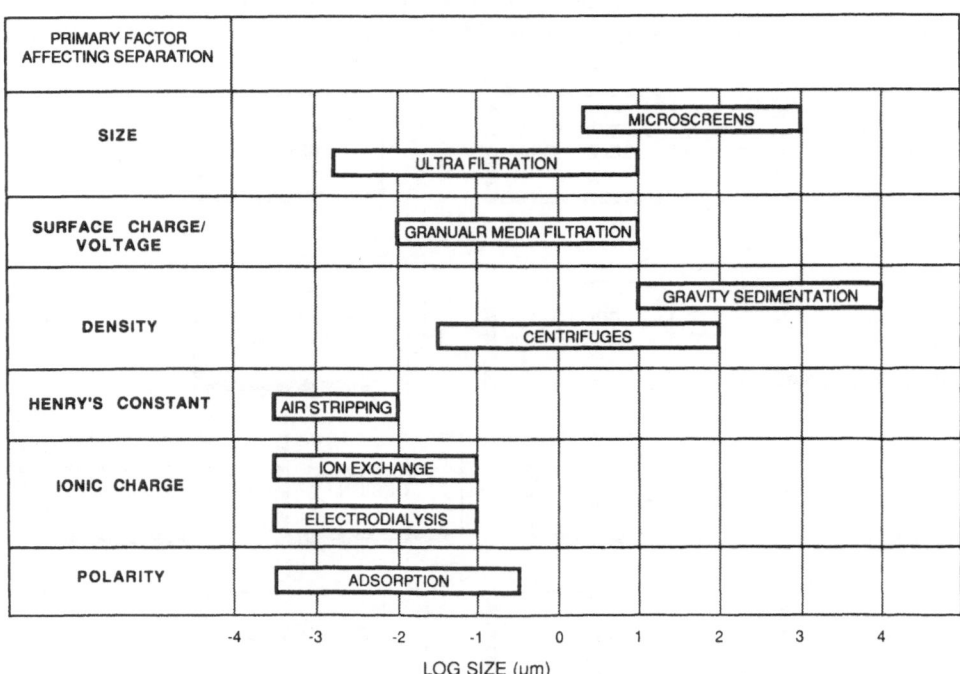

Fig. 2:   Useful ranges of separation processes (modified after a diagram prepared by
           Dorr–Oliver Corporation).

constituent and the desired removal efficiency, different gas liquid contacting processes should be evaluated. As shown in Fig. 4, packed tower aeration appears to cover the widest application ranges and is the process of choice when removals greater than 90 percent are required. Cross flow towers, however, show promise for removal of compounds having low volatility[15] such as ammonia or certain weakly volatile organic compounds such as ethylene dibromide or dibromochloropropane, fumigants commonly found in ground water in California and elsewhere in the United States[3].

These process selection diagrams (Figs. 2-4) can be used in the stage of process analysis dealing with the development of alternatives . Of course they cannot replace the necessary site specific detailed analysis of alternatives. Where possible, it is desirable to use analytical models of individual processes before a final selection of design criteria is made. The following is an assessment of the use of such models for three water treatment processes used for the control of synthetic organic chemicals (SOCs). These include packed tower aeration, adsorption on granulated activated carbon and oxidation with ozone.

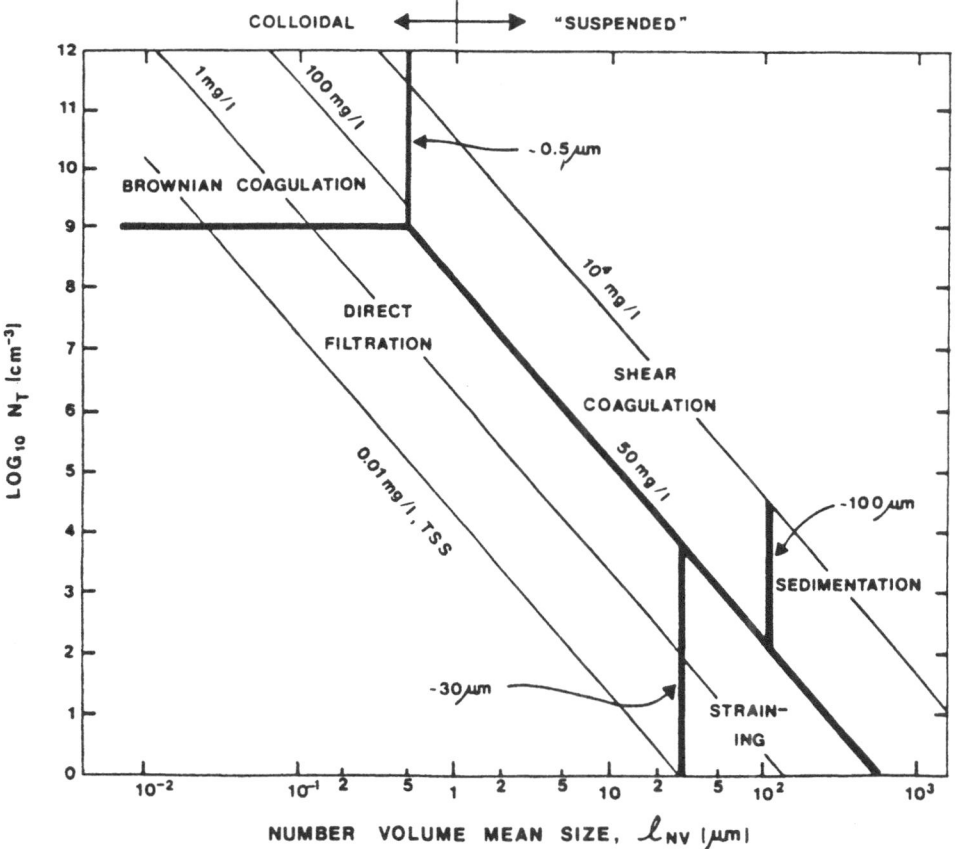

Fig. 3: Solids/liquid separation process selection diagram. Mass concentration isopleths computed based on spherical particles, specific gravity of 1020 kgm^{-3} (from Ref. 10).

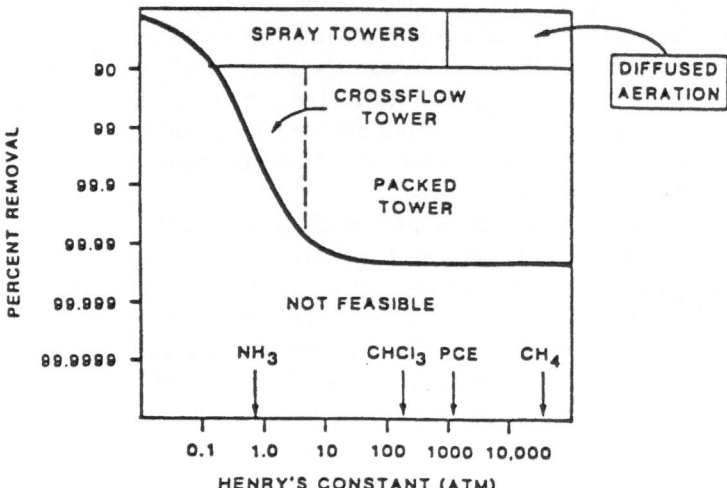

Fig. 4:   Schematic diagram for selection of air stripping processes for removal of
             volatile contaminants (from Ref. 12).

5. Aeration Processes for SOC Control

5.1. Process Description

   Aeration processes can be used as a separation technology for the removal of volatile
constituents transferring the materials from water to the atmosphere. This process can be used
for removal of compounds exhibiting a wide range of volatility. Compounds with low Henry's
constants such as ammonia, can be removed from water by aeration processes if the equipment
can be designed to handle very large air to water ratios. Process performance depends upon
efficient contact between the air and water phases to provide an adequate rate of mass transfer.
Various process options are available for achieving this interphase contact. Water can be
ejected into the air phase (spray towers) or air can be injected into the water (diffused air or
mechanical aeration). As mentioned previously, for removals exceeding 90% packed towers
have been shown to be the most cost effective process choice and will be used here to illustrate
the application of process analysis.

   Transport models have been developed to describe the performance of packed towers[12,15],
using the maximum gradient steady-state model which is based on describing concentration
profiles in a single direction. The general differential equations describing packed tower
performance, including axial dispersion in either the gas or liquid phase, have been developed
by Mecklenburgh and Hartland[14] and Selleck[15] and are given as follows:

$$\frac{d}{dZ}\left(X - \frac{1}{P}\frac{dX}{dZ}\right) = \frac{U_G}{U_L}\left[\frac{d}{dZ}\left(Y + \frac{1}{R}\frac{dY}{dZ}\right)\right] = N(X^* - X) \qquad (1)$$

where X and Y are the bulk concentrations of the constituent in water and air respectively, Z
a dimensionless vertical distance ($Z = Z'/Z$) where Z' is the height in the tower, and Z is the
total packing height, and U_L and U_G are the water and air superficial velocities, with appropriate

units. X^* is the equilibrium concentration in the water defined by Henry's law, given as equation 2:

$$Y = H \cdot X^*$$

(2)

with H as the Henry's constant (dimensionless). Finally, the three dimensionless numbers in equation 1 are N = number of transfer units, and P and R the Peclet numbers for liquid and gas phases, given as $P = U_L Z / E_L$ with E_L an empirical axial dispersion coefficient, and $R = U_G Z / E_G$.

The solution to equation 1 for the packing depth Z is given as[14]:

$$\frac{X_{in} - X_{out}}{X_{out} - Y_{in}/H} = \frac{1 + B}{1 + B/S}$$

(3)

where S = stripping factor = $H Q_G / Q_L$ and B is a function of N, S, P and R.

Assuming no dispersion in either phase, i.e. plug flow behavior, then it can be shown[14] that:

$$B = -\exp\left[N\left(S^{-1} - 1\right)\right]$$

Equation 3 becomes:

$$N = \left(\frac{S}{S-1}\right) \ln \left[\frac{X_{in}/X_{out}(S-1) + 1}{S}\right]$$

(4)

for S > 1.

This is the common form of the equation used to calculate the number of transfer units required in a packed tower for desired removal efficiencies of a specific compound with known Henry's constant.

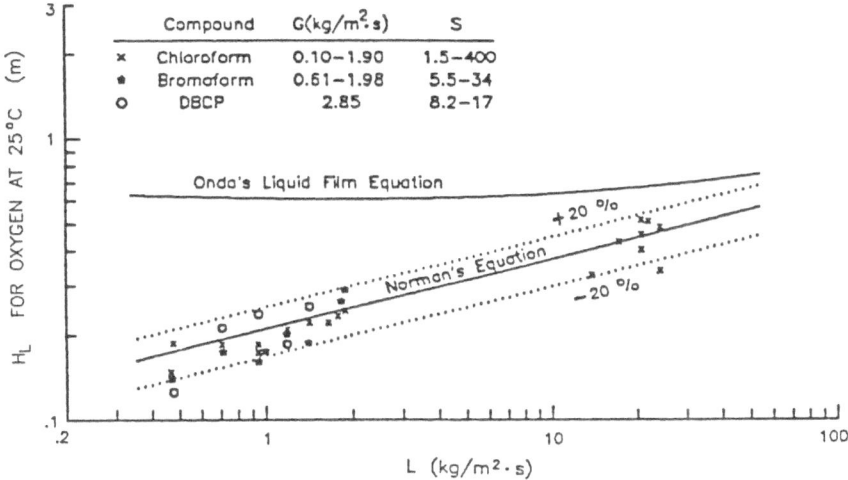

Fig. 5: Comparison of predicted mass transfer coefficients with experimental values for selected organic compounds in water. Shown are predictions of the height of a transfer unit using the Onda correlation[16] and Norman's equation[15].

This model, assuming no dispersion, has been successfully employed in the selection of design criteria for packed towers which have been used to remove of a wide range of compounds[12]. Limitations on the model include the neglect of dispersion which can lead to reduction in performance especially when removal efficiencies greater than 99% are required. The model also assumes steady state performance and does not account for variable influent quality or variable environmental conditions. These variations can be accounted for through sensitivity analysis, thus providing a quantitative basis for determining an overdesign factor.

At the same time, the design problem cannot be resolved without the use of several empirical models both for air pressure drop and for the mass transfer coefficients. Air pressure drop information is usually supplied by equipment and packing manufacturers. Mass transfer coefficients, on the other hand, are more difficult to obtain. Several general correlations have been developed to permit estimates of mass transfer coefficients for different size packings and to account for the effects of air and water flow rates. The two principal correlations are those of Sherwood and Holloway[17] and Onda[16]. Recent work completed by Selleck et al . at the University of California Berkeley[15], indicates that the Sherwood Holloway correlation is the most accurate for most volatile organic carbon compounds (VOCs) of concern as shown in Fig. 5.

The mass transfer coefficient is used directly to determine the height of a transfer unit (HTU) as given by equations. (5) and (6).

$$H_{OL} = H_L + H_G/S \tag{5}$$

Here H_{OL} is the overall transfer unit (dimension of length), H_L is the liquid phase HTU and H_G the gas phase HTU, where:

$$H_{OL} = \frac{U_L}{K_{OL}a} \tag{6}$$

with $K_{OL}a$ the overall mass transfer coefficient (units of $(time)^{-1}$), which can be obtained from the correlations discussed (see Selleck[15] for a summary of correlations used for both H_L and H_G). The overall height of packing is simply the product of H_{OL} and N (see Ref. 12 for description of design procedure).

In summary, stripping of volatile constituents in a packed tower has been adequately described by process models using a transport phenomena based model with maximum gradient description in conjunction with empirical models. This approach permits optimization of the design and quantification of the over-design factor. Where removals greater than 99% are required, pilot testing is still advisable, particularly in situations where unusual water matrixes may be encountered. Continued research on improving packing characteristics would be beneficial. However the air stripping packed tower process is an example of a successful application of process analysis principles leading to selection of optimal design criteria.

6. Adsorption on Granular Activated Carbon

6.1. Process description

A second process alternative for removing synthetic organic chemicals from water is the use of adsorption on granular activated carbon (GAC). In this case the contaminant of concern is transferred from the water phase to a solid surface (GAC) which provides sufficient surface area and active sites for the adsorption process. Process options for contacting the water with the solid phase are more limited compared to air stripping. Fixed bed adsorbers are the most common system utilized although moving bed contacting systems are also available. In the case of fixed bed operations, the process is discontinuous. When the carbon adsorption capacity for a constituent is exhausted, the adsorbent must be disposed of or regenerated. As in the case of the air stripping process, equilibrium data are necessary to determine whether or not the GAC process is suitable for removal of the contaminant.

6.2. General Model

Adsorption in fixed bed adsorbers has been extensively evaluated and the mathematical descriptions are fairly well defined (e.g. see Ref. 18-21). A combined maximum gradient and multiple gradient analysis can be developed to describe concentration profiles in either the water phase or on the solid adsorbent. One such model developed by Weber, Crittenden (Ref. 19 and 20) and others is the homogeneous surface diffusion model (HSDM) which neglects pore diffusion. Using the HSDM, numerical solutions to the following set of differential equations with appropriate boundary conditions, can provide concentration profiles with depth and time. In simplified form, these partial differential equations are given as follows:

For the liquid phase, the concentration of component i is:

$$\frac{\partial C_i}{\partial t} = -U_L \frac{\partial C_i}{\partial Z} + \frac{3\,k_f(1-\varepsilon)}{R\,\phi\,\varepsilon}\left(C_{S_i} - C_i\right) . \tag{7}$$

For the solid phase, a similar mass balance leads to:

$$\frac{\partial q_i}{\partial t} = \frac{D_s}{r^2}\frac{\partial}{\partial r}\left[r^2\frac{\partial q_i}{\partial r}\right] \tag{8}$$

where C_i and q_i are the concentrations of component i in the liquid and solid (adsorbent) phases, Z is the depth, t the time, r the radial distance from the center of the adsorbent particle, ε the void volume in the packed bed, U_L the approach or superficial velocity, k_f the liquid film mass transfer coefficient, D_s, the surface diffusion coefficient, R the particle radius, and ϕ the particle sphericity. Finally C_{S_i} is the equilibrium concentration at the surface of the adsorbent, which can be described using an appropriate equilibrium isotherm. One widely used isotherm model is the empirical Freundlich isotherm, given as:

$$q_i = K\,C_{S_i}^{1/n} \tag{9}$$

for any component i.

In order to solve this set of equations for multi-component adsorption problems, considerable information is required on process parameters. These include the Freundlich isotherm parameters for each contaminant, the film transfer coefficient k_f and the surface diffusion coefficient D_s. Finally, in multicomponent systems, a single component isotherm must be corrected for competitive effects. This can be achieved through the use of the ideal adsorbed solution theory, originally developed by Radke and Prauznitz[22] and expanded by Crittenden and others[20]. The film transfer coefficient k_f can be estimated using the correlation developed by Williamson[23]. Surface diffusion coefficients, D_s, can be obtained from literature values or estimated using a correlation procedure developed by Crittenden[24].

Thus, these transport phenomena based models, with appropriate initial and boundary conditions, combined with empirical models of process parameters provide the engineer with powerful and rapid tools to conduct a detailed process analysis of GAC adsorption for control of SOCs in water. Software is available to use this procedure for such design analysis (see e.g. Ref. 24). Fig. 6 illustrates a recent successful simulation of trichloroethylene (TCE) removal by GAC in a seven component mixture[26], using the more complex pore and surface diffusion model.

Unfortunately the adsorption process is complicated by several phenomena which restrict the accuracy of predictive models. These include the impact of biodegradation and the effect of unknown or uncharacterized constituents on adsorption capacity. It has been demonstrated that naturally occurring organic matter in water can reduce the capacity of the GAC for the removal of trace synthetic organic chemicals. Recent work by Zimmer[25] illustrates the impact of the adsorption of natural organic matter on GAC equilibrium isotherms for other organic compounds present in the water matrix, as shown in Fig. 7. A decrease in the adsorption capacity term, K, can be accounted for by defining an empirical model relating the decrease in K with the loading time as shown in Fig. 8. Crittenden and others[26] have successfully applied this model to the evaluation of field data from a number of different sources. Fig. 9 shows the removal of trichloroethylene by GAC that has adsorbed a known amount of natural organic matter.

Further complications in the use of predictive models of GAC adsorption include the impact of backwashing. This factor has been recently accounted for by Crittenden[26] whereby the impact of backwashing on column performance can be predicted.

In summary, the GAC process is considerably more complex than the air stripping process. Mathematical models have been useful for the assessment of GAC process feasibility for SOC removal in ground and surface waters. These models are currently being utilized by the U.S.

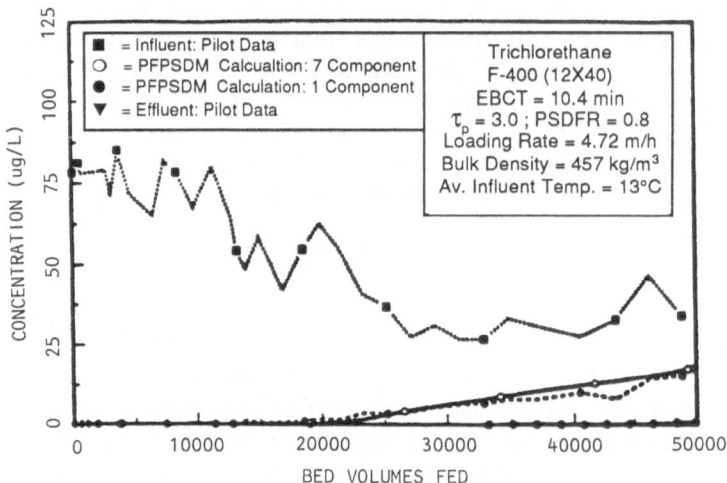

Fig. 6:   Comparison of model predictions using the pore and surface diffusion model (PSDM) for experimental results, 7 component mixture in ground water. Data show TCE effluent history compared to 7 component and single component predictions (from Ref. 26).

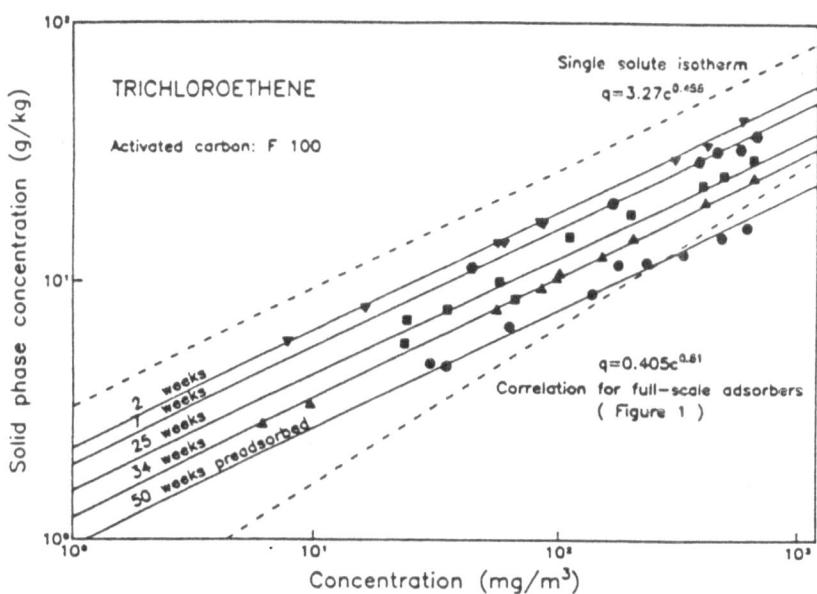

Fig. 7: TCE isotherms for GAC that has been preloaded with 0.7 g/m³ of natural organic matter for various times (after Ref. 25).

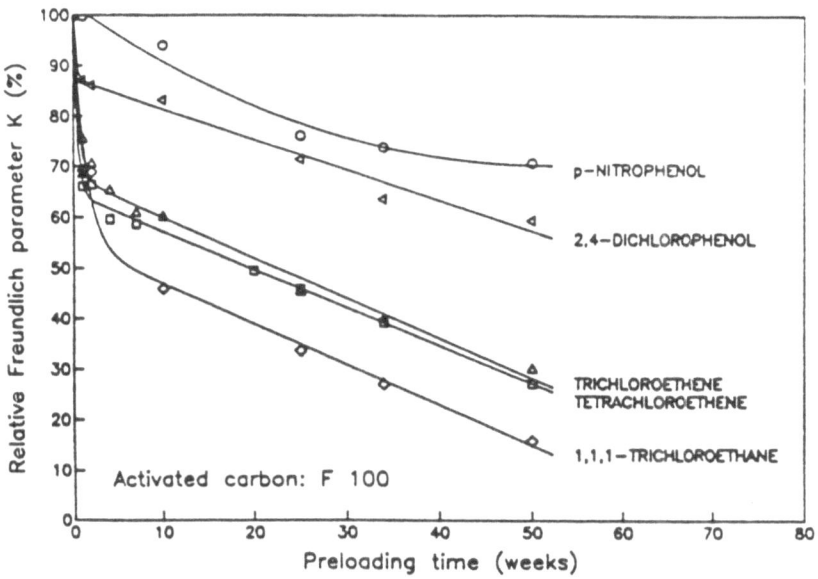

Fig. 8: Empirical description of decrease in Freundlich capacity parameter, K, as function of preloading time, abstracted from Fig. 7 (after Ref. 25).

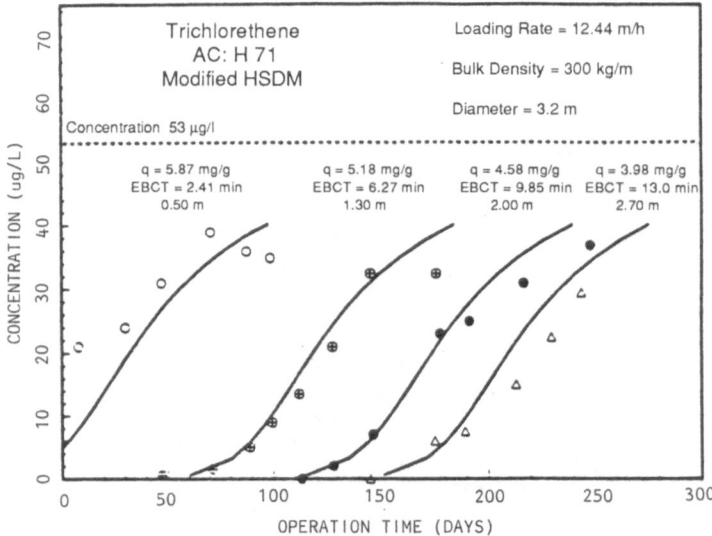

Fig. 9:   Comparison of actual plant data for TCE removal at a German city treating
           ground water containing TCE (53 $\mu g/l$) and PCE (16 $\mu g/l$), with predictions
           using the HSDM and the empirical relation for K (after Ref. 26).

Environmental Protection Agency[27] to determine the suitability of GAC to meet the
requirements of the Safe Drinking Water Act. The process is complex and current process
analysis models are limited because of site specific impact on the adsorption of dissolved
organic matter and the difficult problem of chromatographic effects due to variable influent
quality in the adsorption column. However, current microcomputer models are adequate for
preliminary design purposes. Pilot studies are usually appropriate because of unknown and
unpredictable matrix effects. A combination of mathematical modeling with rapid short
column experiments may be adequate to provide the necessary design data for most separation
problems in potable water treatment[24]. This is not the case for waste water applications
however, where competitive effects and biodegradation phenomena play a dominant role and
are more difficult to quantify.

7. Ozone Oxidation Process

7.1. Process Description

Both the packed tower air stripping and the GAC adsorption processes suffer the significant
disadvantage of transferring a toxic contaminant from one environmental medium to another.
This can potentially create a secondary contamination issue, although air emission control
devices and regeneration of the spent carbon can minimize such problems. A desirable
alternative to these processes would have the ability to convert the toxic or hazardous
substances to innocuous by-products at a cost comparable to the costs associated with air
stripping and GAC. Oxidation processes such as the use of ozone have this potential. Ozone
is a strong oxidant and when added to water in the appropriate type of gas-liquid contact or with

adequate contact time, oxidation reactions that occur may be sufficiently rapid to produce non-toxic by-products. Other process options include the use of multiple oxidants (e.g. H_2O_2 and ozone) and a photolytic oxidation process involving simultaneous use of UV and ozone. For the purpose of this discussion, the analysis will be restricted to the ozone oxidation process.

For gas/liquid oxidation processes, a number of contactor options are available. These include columns with random packing, tray towers or no packing (i.e. bubble columns) and agitated tanks. Each of these contactor options provide certain advantages as summarized in Table 5, based on the work of Levenspiel[28] and Kramer and Westerterp[29]. As shown, if very high interfacial surface area per volume of liquid is desired, the spray, packed or plate columns are preferred. On the other hand, if a high volume fraction of liquid is desired or necessary, an agitated bubble contactor can provide the desired characteristics. For very rapid reactions, as in ozone oxidation processes, a large interfacial surface per volume of liquid is desirable. However, for practical reasons, agitated bubble contactors are more widely used.

7.2. General Model Description

Oxidation processes involve a number of phenomena including mass transfer of the oxidant, often in gas form, into the water and chemical reactions between oxidant species and the contaminants of concern. Recent work by Hoigné and coworkers[30,31] have provided a relatively complete picture of the reaction kinetics of ozone when added to water. While the reaction pathways are complex, it is possible to describe the relative oxidation rates using a simplified model as described below[32,35].

Consider either a plug flow type reactor or an agitated bubble reactor in which ozone and water containing undesirable organic contaminants are mixed. The process objective is removal of the undesirable contaminants and their conversion into non-toxic by-products.

For a plug flow reactor, the maximum gradient model provides prediction of the time required to give any desired removal efficiency provided that the reaction kinetics are

Table 5: Equipment characteristics for gas-liquid contactors[28].

Type of Contactor	Interfacial Surface Area per Volume of Liquid (m^2/m^3)	Interfacial Surface per Volume of Reactor (m^2/m^3)	Volume Fraction of Liquid
Spray Column	1200	60	0.05
Packed Column	1200	100	0.08
Plate Column	1000	150	0.15
Agitated Bubble Contactor	200	200	0.9
Bubble Contactor	20	20	0.98

understood. Thus (see Ref. 28), residence time is:

$$T_{PF} = \int_{C_o}^{a} \frac{dC_i}{-r_i} \tag{10}$$

where C_i is the concentration of the contaminant, and r_i is the rate of reaction.

For a stirred tank (ST) reactor, assuming complete mixing, a mass balance around the reactor for the contaminant gives:

$$T_{ST} = -\frac{(C_o - C)}{(1 - \varepsilon_G) r_i} \tag{11}$$

where ε_G is the volume fraction of gas in the contactor, and C_o is the initial concentration.

To solve both these equations, the reaction kinetics must be known. Ozone can oxidize organic molecules in water either directly or indirectly through the production of hydroxyl radicals. Because direct ozone reactions are so much slower than oxidation with hydroxyl radicals[30,31], this process can be neglected. Thus, the rate of reaction between organic compounds in water and hydroxyl radicals is given by[30] the following second order reaction:

$$r_i = -k_{OH} (C_{OH}) (C_i) \tag{12}$$

where K_{OH} is the rate constant between hydroxyl radicals and contaminant i. Hydroxyl radicals can also react with many organic or inorganic species in water, which compete for the radicals produced when ozone reacts with water. It can easily be shown, given these parallel reactions, that the concentration of hydroxyl radicals at steady state, is given by[32]:

$$(C_{OH})_{SS} = r_{OH} \cdot \frac{1}{\sum_i (k_i S_i)} \tag{13}$$

Table 6:   Case 1: Oxidation of PCE in typical ground water

Conditions: pH   = 8
              DOC   = 2 mg C/l
              HCO_3 = 4 mM
              T   = 20 °C
              Q_L = 2000 l/min
              Residence Time = 30 minutes

Removal Required (%)	Plug Flow Reactor		Stirred Tank Reactor	
	Rate of O_3 Addition (kg/hr)	Ozone Consumed (mg/l)	Rate of O_3 Addition (kg/hr)	Ozone Consumed (mg/l)
67	0.25	6.2	0.50	12.5
90	0.51	12.8	2.23	55.7
99	1.02	25.6	24.5	612.
99.9	1.53	38.4		

where r_{OH} = rate of hydroxyl radical production per volume, and k_i the rate constant between OH and scavenger molecule i of concentration S_i. Combining these equations, one obtains for a plug flow reactor

$$\left[r_{OH}\right]_{PF} = \frac{\Sigma(k_i S_i)}{k_{OH} T_{PF}} \ln \frac{C_0}{C} \tag{14}$$

and for the stirred tank reactor,

$$\left[r_{OH}\right]_{ST} = \frac{\left(\frac{C_0}{C}-1\right)\Sigma(k_i S_i)}{T_{ST}(1-\varepsilon_G)k_{OH}}. \tag{15}$$

Thus, if k_i, S_i, and k_{OH} are known, for any given residence time, or desired removal efficiency, the necessary rate of hydroxyl radical production and thus the rate of ozone consumption can be predicted for each type of reactor.

These maximum gradient models can be used to evaluate the feasibility of ozone oxidation under different matrix conditions. In the first case (Table 6) we consider the oxidation of a common volatile organic chemical found in ground water: perchloroethylene (PCE). Typically ground water contains dissolved organic matter in the range of 2 - 4 mg C/l with an alkalinity of approximately 4 mM. Using these data in conjunction with available reaction kinetic data[30,33] an estimate can be made of the required quantity of ozone consumed to achieve various levels of removal of the compound. A second case of interest is PCE in a waste water matrix. Under these conditions the dissolved organic matter is approximately 10 mg/l with the alkalinity at 8 mM. As shown in Table 6, a plug flow reactor is considerably more efficient than a stirred tank reactor as expected. To achieve a 90% removal of PCE in this typical ground water would require approximately 15 mg/l of ozone consumed. The applied ozone dose would, of course, need to be higher because of less than complete ozone dissolution, plus ozone consumption by other oxidation processes, e.g. direct oxidation. The stirred tank reactor, on the other hand, would require almost 56 mg/l of ozone to be consumed to achieve the same level of removal. In both cases, a 30 minute residence time is assumed.

It can be shown that for any type of reactor, the ozone consumed in mg/l is proportional to a reaction kinetic parameter as shown in the following equations.

For the plug flow reactor, the quantity of ozone consumed is given by:

$$O_3 \text{ consumed (mg/}l) = \frac{4.8 \cdot 10^3 \phi}{\eta} \ln C_0/C \tag{16}$$

where $\phi = (\Sigma k_i \cdot S_i)/k_{OH}$ and η is the stoichiometric yield factor, defined as the moles of hydroxyl radical produced per mole of oxidant consumed. For ozone, $\eta = 0.5$ (Ref. 32). For the stirred tank reactor:

$$\phi O_3 \text{ consumed (mg/}l) = \frac{4.8 \cdot 10^3 \phi}{(1-\varepsilon_G)\eta}\left(\frac{C_0}{C}-1\right). \tag{17}$$

Depending on the type of reactor and the relative reaction rate constants of hydroxyl radicals with the contaminant and scavenger molecules, an estimate can be made of the ozone consumption required to achieve the desired removals. This is summarized in Fig. 10 which shows the applied ozone dose required to achieve given removals, shown as a function of the kinetic parameter.

Typically, contaminated ground water contains between 50 to 200 µg/l of a given volatile organic chemical[3]. In the case of PCE, the USEPA has designated a maximum contaminant

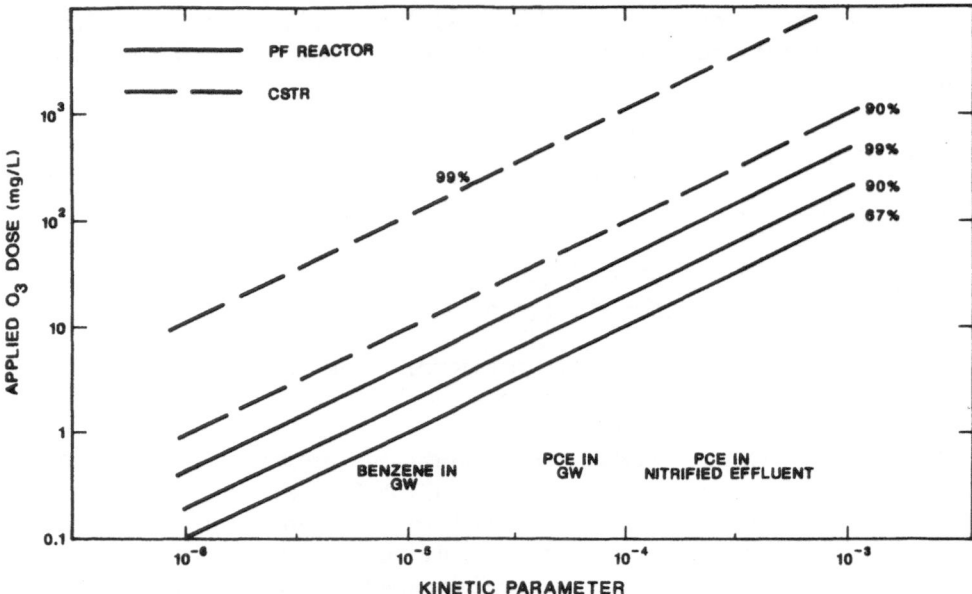

Fig. 10: Estimate of ozone consumption required to achieve various removal efficiencies as a function of relative reaction rates for oxidation of organic compounds in water in the presence of various scavenger molecules. Estimates are shown for plug flow and stirred tank reactors. Examples indicated include benzene and PCE in typical ground water, and PCE in nitrified waste water effluent.

level of 5 $\mu g/l$. In some states, notably California, treatment requirements could be as low as the detection limit (approximately 0.1 $\mu g/l$). Thus, removal percentages in the range of 90 to 99.9% may be required of any oxidation or removal process. For a typical value of the kinetic parameter of 10^{-4}, it can be seen that 90% removal of a trace contaminant requires 22 mg/l of ozone consumed, compared to 44 mg/l for 99% removal. A stirred tank reactor would require substantially more ozone. Clearly the 99% removal efficiency in a single reactor would be infeasible.

Given that the reaction constants are valid only for the initial oxidation reaction and not for the complete oxidation of compounds, the feasibility of using ozonation for complete oxidative removal of trace levels of SOCs is open to question. However, this simplified analysis provides a basis for improving oxidation efficiency. Options include decreasing the concentration of scavenger molecules, i.e. removing the dissolved organic matter and the bicarbonate ions, increasing the stoichiometric yield factor by the use of multiple oxidants such as UV/ozone or hydrogen peroxide/ozone and using a series of stirred tanks to approach plug flow. Recycling of the ozone and increasing the ozone concentration in the feed gas should also lead to an improved rate of oxidation.

In summary, mathematical models of the oxidation process are not yet capable of predicting the fate of synthetic organic chemicals in oxidation reactors. The primary reason for this situation is the unknown reaction kinetics. Additional investigations into reaction kinetics in a variety of water matrices coupled with improvements in gas-liquid contactor design and continued investigation of multiple oxidant systems[36] could lead to an effective process for the complete oxidation of SOCs in ground or surface water.

References

1 James M. Montgomery, Consulting Engineers, Inc., *Water Treatment Principles and Design* (Wiley Interscience, New York, NY.) 1985.

2 Smith, R.A., Alexander, R.B. and Wolman, M.G., Water quality trends in the nations rivers. Science **235** (1987) 1607-1615.

3 Office of Technology Assessment, *Protecting the Nations Groundwater from Contamination*, OTA-0-233, (U.S. Congress, Washington, D.C.) October 1984.

4 H.I. Shuwal (ed.), *Water Renovation and Reuse* (Academic Press, N.Y.) 1977.

5 Baker, M.N., *The Quest for Pure Water* (American Water Works Association, 2nd ed.) 1981.

6 McCarty, P.L. and Reinhard, M., Trace organics removal by advanced wastewater treatment, J. Water Poll. Contr. **52**, 7 (1980) 1907.

7 Rudd, D.F and Watson, C.C., *Strategy of Process Engineering* (John Wiley & Sons, New York, N.Y.) 1968.

8 Himmelblau, D.M., Bischoff, K.B., *Process Analysis and Simulation: Deterministic Systems* (John Wiley & Sons) 1968.

9 National Academy Press, *Toxicity Testing* (Washington, D.C.) 1984.

10 Kavanaugh, M.C., Tate, C.H., Trussell, A.R., Trussell, R.R. and Treewick, G., "Use of particle size distribution measurements for selection and control of solid liquid separation processes" in: *Particulates in Water*, Kavanaugh M.C. and Leckie, J.O., Advances in Chemistry Series, No.189 (American Chemical Society, Washington, D.C.) 1980.

11 Wiesner, M.R., O'Melia, C.R. and Cohen, J.L., Optimal water treatment plant design. J. Environ. Eng. Div., American Soc. of Civ. Eng. **113 EE3** (1987) 567-584.

12 Kavanaugh, M.C. and Trussell, R.R., Design of aeration towers to strip volatile contaminants from drinking water. J. Am. Water Works Assoc. **72**, 12 (1980) 684.

13 Kavanaugh, M.C. and Trussell, R.R., "Air stripping as a treatment process" in *Strategies for Control of Groundwater Contaminants* (American Water Works Association, Denver, CO) 1981

14 Mecklenburg, J.C. and Hartland, S., *The Theory of Back Mixing* (John Wiley & Sons, N.Y.) 1975.

15 Selleck, R.E., The Removal of Volatile Organic Compounds from Water with Aeration Columns Generic Background, Report (prepared for Lawrence Livermore National Laboratory, Livermore, CA and Cal. Dep. of Health Services, Berkeley, CA) December 1986.

16 Onda, K., et al., Mass transfer coefficients between gas and liquid phases and packed columns. J. Chem. Eng., Japan **1**, 1 (1968) 56-62.

17 Sherwood, T.K. and Holloway, F., Performance of packed towers - Liquid film data for several packings. Trans. Am. Inst. Chem. Eng. **36** (1940) 39-70.

18 Sontheimer, H., et al., *Adsorptionsverfahren zur Wasserreinigung* (Engler-Bunte-Institut der Universität Karlsruhe) 1985.

19 Crittenden, J.C. and Weber, W.J., Jr., A predictive model for a design of fixed-bed adsorbers: Model development and parameter estimation. J. Environ. Eng. Div., Am. Soc. Civ. Eng. **104** (1978) 185.

20 Crittenden, J.C., et al., Design of fixed beds to remove multi component mixtures of volatile organic chemicals. (Presented at the AWWA meeting, Washington, D.C., June 23, 1985.)

21 Thacker, W.E., Crittenden J.C. and Snoeyink, V.L., Modeling of adsorber performance: Variable influent concentration and comparison of adsorbents. J. Wat. Poll. Cont. Fed. **56** (1984) 243.

22 Radke, C.J. and Prausnitz, J.M., Thermodynamics of multi solute adsorption from solute liquid solutions. Am. Inst. Chem. Eng. Journ. **18** (1972) 761.

23 Williamson, J., Bazaire, K. and Geankoplis, C., Liquid phase mass transfer at low Reynolds numbers. Indust. Eng. Chem. Fundam. **2** (1963) 126.

24 Crittenden, J.C., Hand, D.W., Arora, H. and Lykins, B.W., Jr., Design considerations for GAC treatment of organic chemicals. J. AWWA **79**, 1 (1987) 74-81.

25 Zimmer, G., Haist, B. and Sontheimer, H., "The influence of preadsorption of organic matter on the adsorption behavior of chlorinated hydrocarbons" in *Proc. of AWWA National Conference*, Kansas City, MO, June 14-

18, 1987.

26 Crittenden, J.C., Hand, D.W., Arora, H., Miller, J., Zimmer, G. and Sontheimer, H., "Adequacy of mathematical models for designing full scale adsorption facilites which remove synthetic organic chemicals" in *Proc. of AWWA National Conference*, Kansas City, MO, June 14, 1987.

27 Adams, J.Q., Clark, R.M. and Miltner, R.J., Cost and performance evaluation for full scale single solute control of synthetic organic chemicals by granular brand new activated carbon adsorptions. (Presented at the AWWA National Conference, Kansas City, MO, June 11, 1987.)

28 Levenspiel, O., *Chemical Reaction Engineering*, 2nd ed., (John Wiley & Sons, N.Y.) 1972.

29 Kramer, H. and Westerterp, K.R., *Elements of Chemical Reactor Design and Operation* (Academic Press, N.Y.) 1963.

30 Hoigné, J. and Bader, H., Role of hydroxyl radical reactions in ozonation processes in aqueous solutions. Wat. Res. **10** (1976) 377-386.

31 Staehelin, J. and Hoigné, J., Decomposition of ozone in water in the presence of organic solutes acting as promoters and inhibitors of radical chain reactions. Envir. Sci. Tech. **19**, 12 (1985) 1206.

32 Hoigné, J., Bader, H. and Knowle, L., Rate constants for OH radical scavenging by humic substances: Role in ozonation and in a few photochemical processes for the elimination of micropollutants. (Paper presented before the Division of Environm. Chemistry, Am. Chemical Society, Denver, CO, April 5-10, 1987.)

33 Farhataziz and Ross, A.B., *Selected Specific Rates of Reactions of Transients from Water in Aqueous Solution: III. Hydroxyl Radical and Perhydroxyl Radical and their Radical Ions* (National Bureau of Standards, Nat. Stand. Ref. Data Ser., NBS-59) Jan. 1977.

34 Hoigné, J. and Bader, H., Rate constants of reactions of ozone and organic/inorganic compounds in water - 1: Nondissociating organic compounds. Wat. Res. **17** (1983) 173-183.

35 Hoigné, J. and Bader, H., Ozone intitiated oxidations of solutes in wastewater: A reaction kinetic approach. Prog. Wat. Tech. **10**, 5/6 (1978) 657.

36 Glaze, W.H., et al., *Oxidation of Water Supply Retractory Species by Ozone with Ultraviolet Radiation.* Report published for the Municipal Environmental Research Laboratory (USEPA, Report No. EPA-600/2-80-110) August 1980.

Discussion

Chairman: H. Strathmann
Fraunhofer Institut, Stuttgart, West Germany

M. Campagna • You showed a list of the compounds being regulated by the US authorities. Do you expect this list to continue to expand at the same rate as it has in the last ten years? I wonder whether, if this is the development, this would imply major changes in the design of plants and processes. Finally I was wondering what kind of major research areas would you consider to be of primary importance if this is the case?

M.C. Kavanaugh • The question addresses the issues of constituents that are being regulated at least in the US. As I mentioned, 49 are currently being regulated and I estimate possibly as many as a hundred by 1990. The environmental protection agency (EPA) is evaluating 83 additional compounds that might be added to the 49 that are currently being controlled. The next phase of regulatory development with respect to drinking water will address oxidation by-products for disinfection. It is probable that trihalomethane levels in the United States will be reduced from a hundred ppb to some other value as yet to be determined. And the impact of that, of course, is that oxidants such as free chlorine will have to be reduced and alternatives will have to be evaluated. In Europe the focus has, for many years, been on eliminating chlorine, primarily for taste reasons not so much for health reasons. So the trend is clearly towards the elimination of chlorine and the use of other oxidants. — The last part of your question was about what sort of research areas are important for the future. I think certainly the area of improved oxidants for disinfection and the removal of synthetic organic chemicals at low levels. I think oxidation is indeed very important for the future of water treatment, both for municipalities as well as for industrial treatment. I think there will be additional research in the area of by-product formation. Certainly we don't know all we need to know about ozone and some of the other oxidants as well.

W. Lorch • In his introduction, Mr. Kavanaugh said that he was giving an overview of the requirements and in particular research requirements. I think two spheres have been totally omitted. The requirements of the developing world and the requirements of medicine, particularly in the health aspects of the developing world. Perhaps Mr. Kavanaugh would like to comment on these two points.

M.C. Kavanaugh • Well, I appreciate that comment very much. I am not that familiar with the medical side of water treatment but I am familiar with the problems of the developing countries. I spent two years in the peace corps in Guatemala and I am very concerned about water treatment applications and the elimination of acute diseases around the world. However, as you can appreciate, I had to make choices in presenting this snapshot and I chose not to address the developing nations' concerns and problems. But I totally agree that these two areas are extremely important.

S. Stucki • Air stripping of chlorinated compounds is unsatisfactory, as you pointed out, because it is only shifting the problem from the water to the air. There is an alternative to oxidation, namely biological treatment with bacteria that are able to destroy chlorinated compounds. What is your opinion of these processes? What are the chances that they will become competitive with oxidation?

M.C. Kavanaugh •   I am not a specialist in the area of microbiology and the application of genetic engineering or other techniques for the removal of these compounds. The results that are coming out of laboratories around the world though are quite promising for aerobic oxidation of aromatic compounds. It does appear that, with proper control of the nutrients and the matrix, oxidation of some of the non-chlorinated compounds can be achieved. With respect to the chlorinated compounds, results are less promising. Anaerobic degradation of the compounds can be achieved; however, it's a reductive dehalogenation process. For example, anaerobic degradation of perchloroethylene (PCE) leads ultimately to vinyl chloride which does not appear to be that degradable and, of course, is more toxic than PCE. There are research projects ongoing to evaluate in-situ biodegradation using a primary substrate of methane or propane, thereby leading to secondary degradation of these compounds. These show some promise. But again, it does not seem that the required level of degradation (above 99%) is going to be easily achievable. This is a very active area of research and one that I probably should have addressed in my overview, but I chose to focus primarily on physio-chemical treatment.

W.T. Rippin •   You referred in your talk to the importance of quantitative overdesign. Would you care to indicate how you approach that problem?

M.C. Kavanaugh •   In the air stripping area we approach a quantitative assessment of the overdesign factor by using the model and stretching it with respect to percent ranges of the parameters of interest: the Henry's constant, mass transfer coefficients and also the influent concentrations. And in that case of course the model is relatively straightforward and can certainly easily be evaluated. This can also be done with activated carbon treatment. The problem there is that an infinite combination of possible influent qualities makes predictive calculations quite difficult because of chromatographic effects. In the air stripping process, we think we have a pretty good handle on that for a quantitive specification of the overdesign factor. In the GAC area I think we are still in an empirical mode.

H. Strathmann •   The traditional approach to waste water treatment was to centralize the water streams, according to the slogan: "Dilution is the solution of pollution". Shouldn't it be the other way round, i.e. to treat the waste water, especially from industrial processes, when it is not diluted?

M.C. Kavanaugh •   I completely agree with that. What we are often dealing with in the areas that I've discussed (the ground water remediation area particularly), is a release to the environment that was not planned for. Immediately there is a very dilute solution, which is highly undesirable of course. The problem is clearly one of dealing with the dilution and still being able to achieve a cost effective solution.

H. Strathmann •   My second question relates to the processes like activated carbon and oxidation. Activated carbon may act as a catalyst and you get products which are even more hazardous than the primary pollutant. The same is true for oxidation, even if you replace chlorine by ozone. What is your opinion?

M.C. Kavanaugh •   I think a very important research area should be in the area of oxidation by-products, proving and showing conclusively what by-products in fact these compounds are oxidized to.

Reverse Osmosis for the Separation of Organics from Aqueous Solutions

R. Rautenbach and I. Janisch
RWTH Aachen, Aachen, West-Germany

1. Introduction

Reverse osmosis (RO) has proved reliable and economically attractive in sea- and brackish water desalination and in the production of pure water. To a large extent this success has to be credited to the new polymer membranes which have been commercially introduced during the last 10 years. Although primarily developed for seawater desalination, their application in the treatment of industrial effluents is tempting because of their separation potential for many organic chemicals. With respect to industrial effluents, this quality is of central importance since such effluents often contain organics as solvents. Reverse osmosis is very suitable for the treatment of industrial effluents since:

- Its modular design allows integration into the individual production process and, consequently, adaption to each individual separation problem;

- The chemicals are separated without any thermal or chemical degradation, thus allowing recovery, at least in principle.

In this paper, the potential of RO for the treatment of organics-contaminated waste water is shown by describing a few interesting cases. First, however, the properties of modern polymer RO membranes with respect to aqueous solutions of organics are discussed.

2. Properties of Polymer RO Membranes with Respect to Aqueous-Organic Systems

In Fig. 1. the rejection coefficient of some composite membranes is plotted against the molecular weight of a number of organic chemicals[1]. The tested chemicals are listed in Table 1. According to Fig. 1, organics with molecular masses M > 100 g/mole are rejected almost totally by all of these membranes.

For molecular weights M < 100 g/mole, membrane selectivity is determined by the molecular structure of the chemicals and by the membrane material itself. The rejection coefficient for acetic acid, for example, of the membrane NTR 7197 from Nitto is only 34%, whereas the rejection coefficient of the membrane PEC 1000 from Toray is 91%.

Here, a short discussion of the dubious, though practical, measure "rejection coefficient" seems appropriate. The rejection coefficient is commonly used in sea- and brackish water

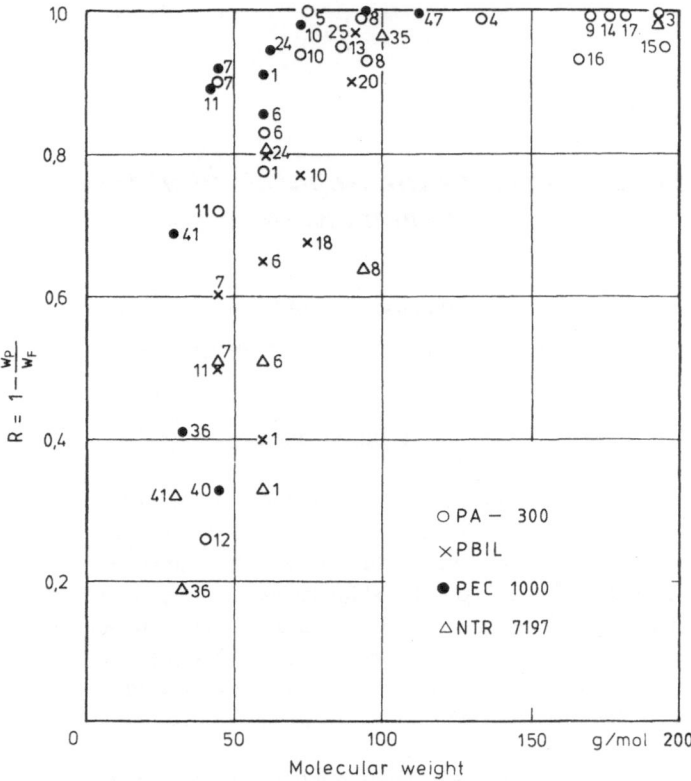

Fig. 1:   Influence of molecular weight on selectivity (rejection). Numbers refer
to the list of chemicals in Table 1.

Table 1:   Chemicals tested in the experiments shown in Fig. 1.

No.	Compound	No.	Compound
1	acetic acid	15	dimethyl phtalate
2	citric acid	16	tetrachloro ethylene
4	D, L-aspartic acid	17	trichloro benzene
5	glycine	18	propionic acid
6	urea	20	oxalic acid
7	ethanol	24	ethylene glycol
8	phenol	25	glycerin
9	o-phenyl phenol	35	N-methyl pyrrolidone
10	methyl ethyl ketone	36	methanol
11	acetaldehyde	40	formic acid
12	acetonitrile	41	formaldehyde
13	ethylaceto acetate	47	ε-caprolactam
14	butyl benzoate		

desalination. It is defined according to

$$R = 1 - w_p/w_F \tag{1}$$

with w_p as the concentration (mass-content) of the locally produced permeate and w_F as the concentration of the local feed at the membrane surface. Even neglecting concentration polarization, i.e. the influence of hydrodynamics on R, the rejection coefficient is not a solute and membrane specific property but is highly influenced by the driving force, i.e. the trans-membrane pressure difference and by the concentration of the feed-side solution. Consequently, a rejection coefficient is only meaningful in connection with precise information about these two figures. Since the trans-membrane pressure difference and the feedside concentration are important and since the aim of the process "RO" is the enrichment of a solution, the rejection coefficient for retentate conditions (high concentrations of organics) will differ from the rejection coefficients for feed conditions (low concentrations of organics).

This is shown in Fig. 2, where the rejection coefficient of butanol solutions in water is plotted against the feedside concentration[2]. With increasing butanol content in the feed, the rejection coefficient decreases drastically and even negative rejection coefficients are observed for low-selectivity membranes.

The characterization of the membrane selectivity by a selectivity curve as shown in Fig. 3 is certainly preferable in such cases.

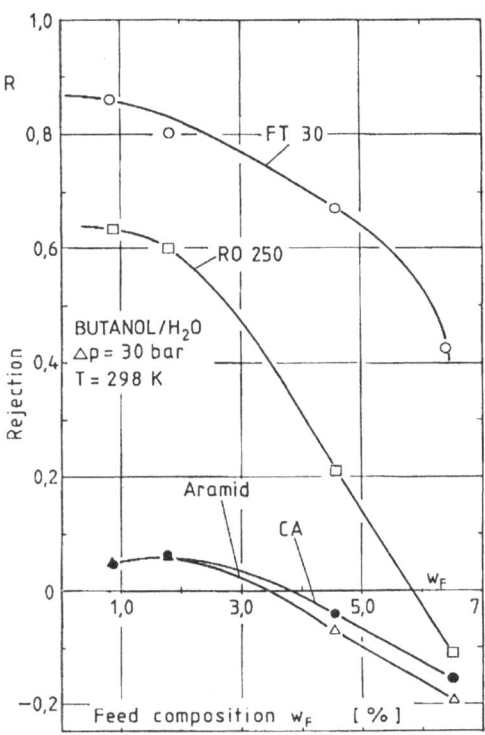

Fig. 2:   Influence of feedside concentration on rejection[2].

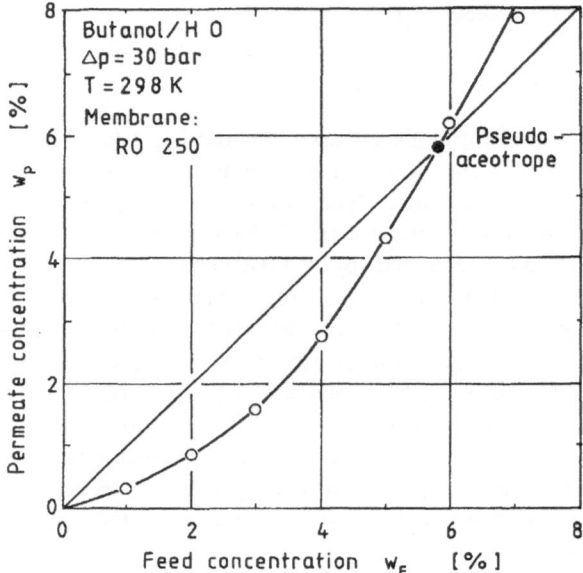

Fig. 3:   Selectivity of the membrane RO 250 for the system butanol/H₂O.

The data shown in Fig. 1 and in Table 2 are valid for aqueous solutions with 0.5 wt% organic content and for a trans-membrane pressure difference of $\Delta p = 30$ bar.

The importance of the molecular structure of the organic chemical for the selectivity of RO membranes is indicated by the experimental results listed in Table 2[3] and Table 3[4].

As a general rule, the rejection of branched-chain molecules is always better than that of straight-chain molecules. Acids are rejected better in their dissociated form than in their undissociated form.

Since the degree of dissociation of weak acids like boric acid or organic acids depends on pH (see eqs. (2) and (3)):

$$\alpha = \left[1 + \frac{10^{-pH}}{K} \right]^{-1} \tag{2}$$

$$K = \frac{[C_6H_5O^-][H^+]}{[C_6H_5OH]}, \tag{3}$$

it must be concluded that membrane selectivity for weak acids also depends on pH. This interdependence is shown in Fig. 4[4], where the degree of dissociation α and the rejection coefficient R are plotted against the pH-value. Both functions follow the same pattern, at least qualitatively.

Besides selectivity, two other membrane properties are important for an economically successful application:

•   chemical stability and

•   permeate flux.

Table 2:   Influence of the structure of molecules with comparable molecular
            weights on selectivity of RO membranes[3].

Component	Molecular Weight [g/mole]	R	Membrane NTR-7197 (Nitto)
iso-propanol	60	0.90	
ethylene glycol	62	0.81	$\Delta p = 30$ bar
acetic acid	60	0.34	$w_F = 5000$ ppm
urea	60	0.52	$T = 293$ K

Table 3:   Effect of branching (comparable molecular weights) on selectivity of RO
            membranes[4].

Component	R	Membrane Du Pont Aramid
n-butanol	0.65	
sec-butanol	0.77	$\Delta p = 28$ bar
iso-butanol	0.95	$w_F = 5000$ ppm
tert-butanol	0.96	$T = 293$ K

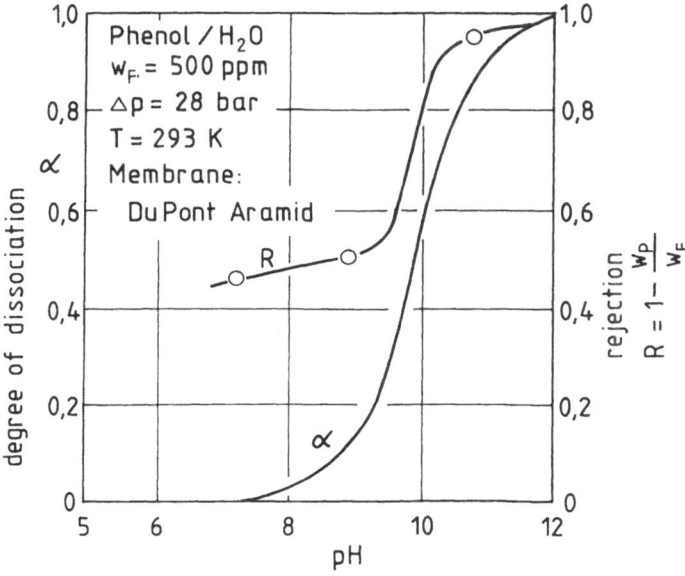

Fig. 4:   Influence of pH on degree of dissociation and on rejection.

Table 4:   Membrane damage by vinyl acetate; standard NaCl test before and after exposure.

Membrane	before		after		
	v_w [l/m²h]	R	v_w [l/m²h]	R	NaCl/H₂O
FT 30 (Filmtec)	52.3	0.97	28.7	0.30	w_F = 5000 ppm
RO 250 (Rochem)	39.0	0.99	105.6	0.03	Δp = 30 bar
Aramid (DuPont)	40.2	0.95	99.2	0.21	T = 298 K
PA 300 (UOP)	50.6	0.93	3.4	0.03	

Although flux is important in process economics, it has been ranked below selectivity and stability because a low flux can be compensated for by additional membrane surface area, at least in principle, whereas selectivity and membrane stability are essential for the process.

Polymer membranes, especially the composite membranes, are much more resistant to organics than cellulose acetate membranes. There are, however, certain chemicals which damage these membranes.

An exposure to a 2 wt% aqueous solution of vinyl acetate for example, resulted after three weeks in irreversible membrane damage. Table 4 shows the results of a standard test with a NaCl solution. The rejection coefficients of the tested RO membranes were drastically reduced after exposure to vinyl acetate. The different changes of flux should be noted.

Whereas selectivity dropped for all membranes, membrane damage resulted in a flux decrease in two cases and a flux increase in the case of the other two membranes. According to our experiments, all acetates like ethyl- and butyl acetate appear to damage RO membranes, the extent depending, however, on concentration. In the case of aqueous-organic solutions, the flux of RO cannot be predicted by the simple relationship approximating the behaviour of salt solutions.

With some very simplifying assumptions, equations 4a and 4b can be derived from the solution diffusion model:

$$v_w \sim v_p = A(\Delta p - \Delta \pi) \tag{4a}$$

or:

$$\frac{v_p}{v_{po}} = 1 - \frac{\Delta \pi}{\Delta p}, \tag{4b}$$

with:

$$v_{po} = v_{wo} = A \Delta p \tag{5}$$

as flux in the case of pure water. For very low concentrations, the osmotic pressure is negligibly small and consequently the flux should be equal to the pure water flux. In reality, however, a

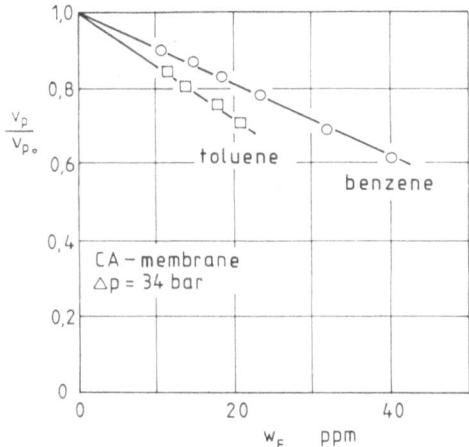

Fig. 5: Decrease of flux in RO caused by traces of organics[5].

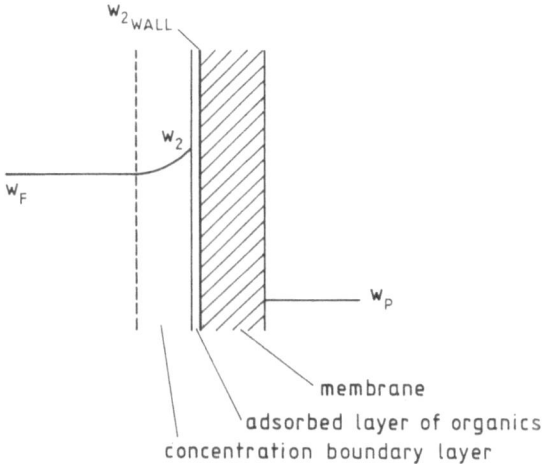

Fig. 6: Concentration distribution of the organic chemical in RO of aqueous-organic solutions.

different behaviour can be observed. Even traces of organics lead to a marked flux decrease (Fig. 5.)[5].

A possible explanation for these, compared with eq. (4) lower fluxes, could be the existence of an adsorbed layer of the rejected organic component at the membrane surface (Fig. 6). Though in principle eq. (4) could still be valid, it is useless as long as the degree of adsorption and, in turn, the corresponding real osmotic pressure at the membrane surface are not known.

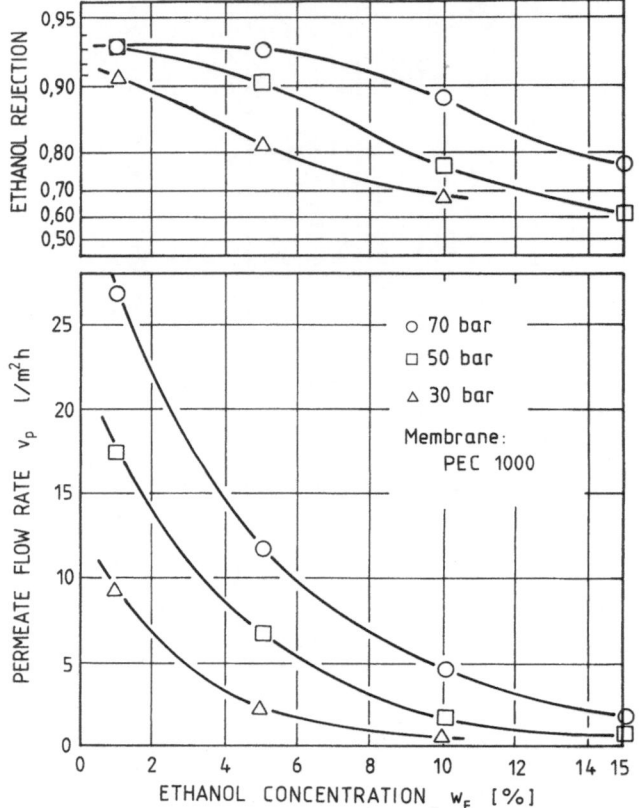

Fig. 7:   Selectivity and flux of the PEC 1000 composite membrane (TORAY) for
               ethanol-water solutions[6].

For calculation purposes, Dickson and Lloyd[5] propose an empirical relationship of the form:

$$\frac{v_p}{v_{po}} = 1 - K\, x^n_{org, F} \tag{6}$$

Since a sound, i.e. generally valid, mathematical relationship for mass transport in RO
membranes for separating organics from aqueous solutions does not yet exist, membrane
manufacturers provide information on flux and selectivity in the form of diagrams as shown
in Fig. 7 for ethanol, H_2O and for the PEC 1000 composite membrane from Toray[6].

Such detailed information is, however, limited to a very few chemicals and membranes. This
is the reason why our institute has started a large experimental program on RO (better: pressure
permeation) of binary and ternary aqueous-organic mixtures.

3. Applications

3.1. Landfill Drainage Water

Drainage water from landfills can be highly polluted by inorganic and organic components as shown in Table 5 for some examples. In Table 5, the chemical oxygen demand COD is listed as a measure of the organic content and the electrical conductivity EC as a measure of the inorganic content.

Usually, most of the organic components of landfill drainage water are low molecular weight chemicals such as solvents (Fig. 8).[7]

All the drainage effluents listed in Table 5 have been subjected to RO tests (Table 6). According to these tests, the selectivity of reverse osmosis is generally sufficient for the permeate to be discharged into the municipal sewage treatment plant or even into a river. In cases where an even better permeate quality is required, a two-stage RO has to be installed.

Table 5:   Organic and inorganic pollution of landfill drainage water.

No.	COD [mg/l]	Conductivity EC [mS/m]	Source
1	6127	3740	Ref. 7
2	2480	1980	Ref. 8
3	21950	—	Ref. 9
4	26400	1780	Ref.10

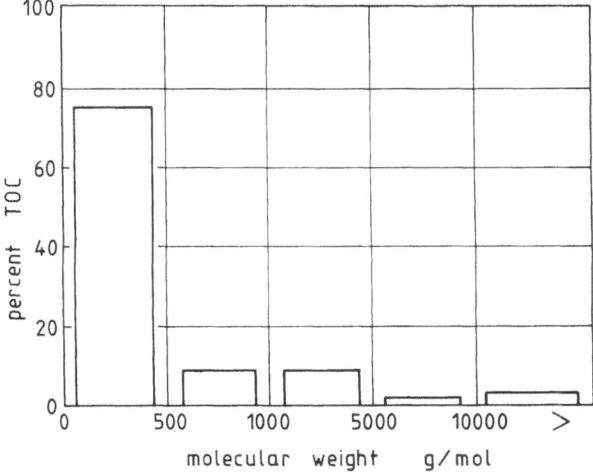

Fig. 8:   Molecular weight distribution of organics contained in landfill drainage water[11].

Fig. 9:   RO plant for treatment of landfill drainage water, equipped with tubular
                 modules and sponge ball cleaning[12].

Table 6:   Selectivity of RO for landfill drainage water.

No.	$R_{COD} = 1 - \dfrac{COD_P}{COD_F}$	$R_{EC} = 1 - \dfrac{EC_P}{EC_F}$	Membrane	Conditions
1	0.789	0.742	FT 30	—
2	0.982	0.982	FT 30	$\Delta p = 40$ bar, T = 20 °C
3	0.989	—	—	—
4	0.675 (0.650)	0.950 (0.910)	CA	recovery: 0% (40%) $\Delta p = 28$ bar, T = 21 °C

With respect to landfill drainage water, neither selectivity nor membrane stability are critical. One problem is, however, membrane fouling and, as a consequence, the rather low permeate flux. Furthermore, the disposal of the RO concentrate can present a problem.

Fouling can be controlled by the installation of tubular modules in combination with sponge ball cleaning. This has been demonstrated by the RO plant of Hager and Elsässer (Fig. 9)[12] installed at the municipal landfill "Hinterer Dollert" near Rastatt, West Germany. The plant has been operating successfully without any feedwater pretreatment since September 1986 and without any membrane damage caused by the sponge ball cleaning.

In this particular case, the RO concentrate is recycled to the landfill. This should be possible in other applications since RO generally controls only the water balance of the landfill.

The plant operates at a recovery rate of 75%, the specific treatment costs are about 13 DM/m³ permeate. In cases where the RO concentrate cannot be recycled, (for example, in a landfill without a bottom sealing), the RO retentate has to be concentrated further by evaporation or combustion. This is an expensive process, but certainly less expensive than evaporation alone since such a combination means that the evaporation stage has only to be designed for the relatively small volume of the RO concentrate (see section 3.3.).

3.2. Desorbate (Steam-Stripping Condensate)

Adsorption on active charcoal is one of the most commonly used techniques for the separation of organics from water or air. The active charcoal is arranged in the form of fixed beds which are regenerated (desorbed) periodically by steam-stripping. In many actual applications, the vapor from this regenerating step – a mixture of organics and steam – forms two phases after total condensation because of a partial immiscibility. The organic phase, containing only small amounts of water, can be reused if necessary after a further purification. The water phase, however, can be a problem since its organic content is definitely too high to be discharged into the sewage system. As long as adsorption is used in the treatment of contaminated water, e.g. ground water contaminated by chlorinated hydrocarbons[13], this water phase from the phase separation can be recycled to the adsorber feed (Fig. 10).

In the case of air cleaning by adsorption, however, this desorbate has to be treated further. An example is the adsorption for the cleaning of 500,000 m^3/d exhaust air from paint-drying ovens. The solvent concentration of about 2000 mg/m^3 is reduced by this step to 20 mg/m^3, i.e. about 1000 kg of solvents are separated per day. After phase separation of the steam-stripping condensate, an organic phase containing about 98 wt% solvents and a water phase containing about 3.7 wt% are obtained. Since the solvents in this desorbate water have a wide range of boiling points and since some components form aceotropes with water, separation by distillation is complex and expensive.

As an alternative, a hybrid process consisting of a RO stage, a second phase separation and a pervaporation (PV) stage has been designed and tested in a pilot-plant[14]. Fig. 11 shows the flow diagram of the process. Table 7 lists the main process streams and their compositions.

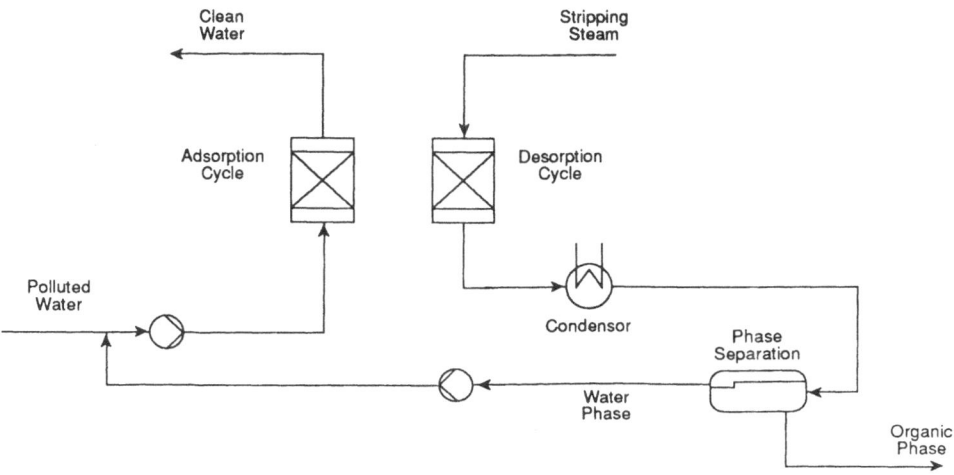

Fig. 10: Separation of organics from groundwater contaminated by chlorinated hydrocarbons[13].

Table 7:   Material balance of the hybrid process.

Stream	Mass Flow Rate [kg/d]	Composition [wt% H_2O]
1	7000	100.0
2	8000	87.5
3	750	2.0
4	7250	96.3
5	17372	91.5
6	6998	99.7
7	10374	86.0
8	357	30.0
9	10017	88.0
10	252	2.0
11	105	97.0

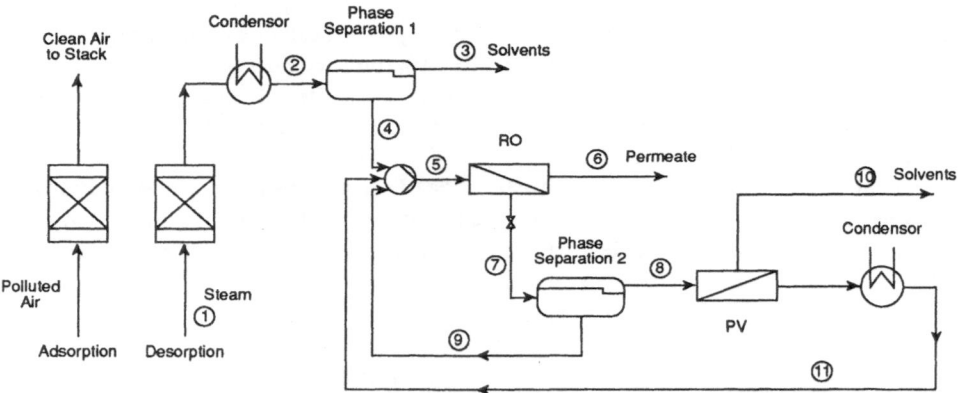

Fig. 11:   Flow diagram of exhaust air cleaning by adsorption, including treatment of the steam-stripping condensate by membrane processes[14].

Such a process is feasible since RO separates water from the equilibrium system "desorbate water", thus inducing a further phase separation of the RO-concentrate. The organic phase of this second phase separation contains about 36 wt% water and is purified by pervaporation to about 2 wt% water content for further use.

The permeate of the pervaporation, containing about 4 wt% organics, is recycled to the RO–feed. It should be emphasized that the combination of phase separation and reverse osmosis operates at a process recovery rate of 90% (process feed concentration 3.7 wt% organics) without exceeding 10 ÷ 12 wt% organics concentration in the RO modules!

According to our tests, several composite membranes, for example the Filmtec FT 30, the Toray PEC 1000 and the UOP TFC are suitable. Spiral wound modules equipped with FT 30 membranes have been in operation for more than 1000 hours without any noticeable changes

Table 8:   Separation of water from desorbate water by RO, experimental results[15].

Component	Feed-Comp. w_F [g/l]	$R = 1 - \dfrac{w_P}{W_F}$	FT - 30 (FilmTec)
iso-propanol	5.859	0.820	Spiral-wound module
iso-butanol	7.133	0.903	$\Delta p = 60$ bar
n-butanol	40.530	0.860	
diglycol-mono-butylether	62.409	0.979	T = 298 K

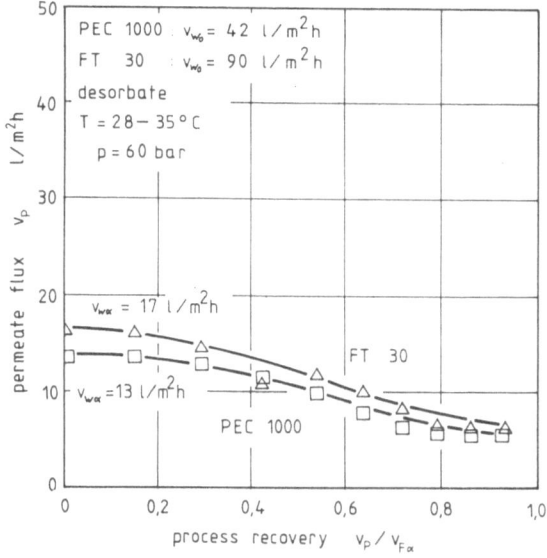

Fig. 12:   Influence of process recovery (RO-feed concentration) on permeate flux[15].

in performance. In Table 8, some results achieved with the FT 30 membranes at a recovery rate of the RO stage of about 16% and a feed concentration of about 10 wt% are listed[15].

According to Fig. 12, the permeate flux is rather low for all membranes, decreasing from about 15 l/m^2 h for a process recovery rate of zero, corresponding to a RO-feed concentration of 3.7 wt%, to about 5 l/m^2 h for a process recovery rate of 90%, corresponding to a RO-feed concentration of about 10 wt% organics[15].

These low fluxes are most probably caused by an adsorbed layer of organics on the

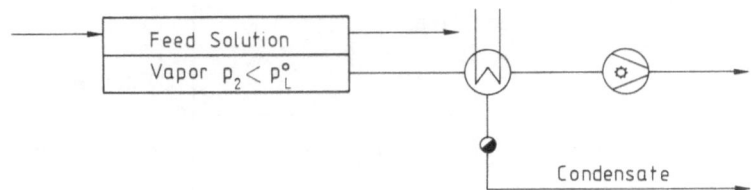

Fig. 13:   Principle of pervaporation.

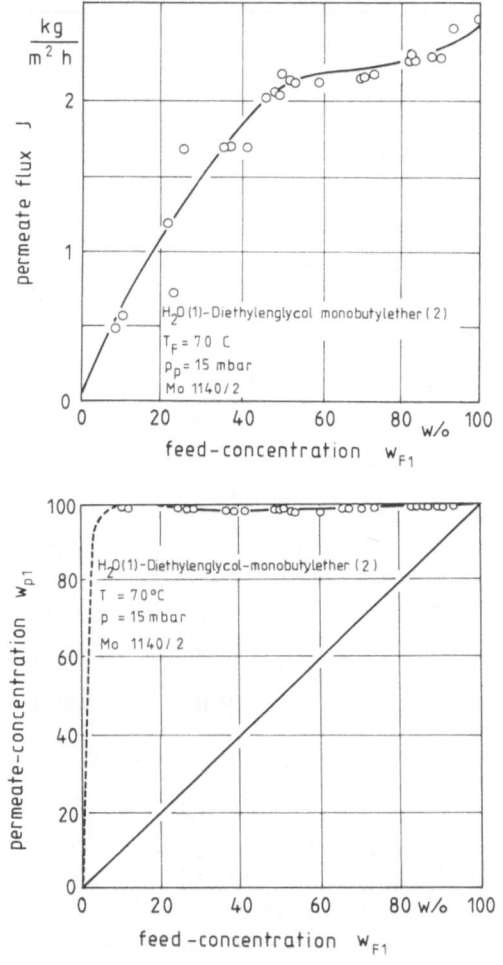

Fig. 14:   Flux and selectivity of pervaporation[14].

membrane surface as discussed in section 2 (Fig. 6). The low fluxes (i.e. high specific membrane area) in combination with the low recovery rate of the RO modules (i.e. high cross flow rates) result in the relatively high specific costs of this stage. It should be noted that the separation loop includes the phase separation (Fig. 11). Whereas in the RO plant for the treatment of landfill drainage water the high flow rates, which are characteristic of tubular systems, have to be maintained only within the high pressure loop; in the case of desorbate treatment the high flow rate stream (stream 7 in Fig. 11) have to be pressurized and then depressurized for the phase separation.

Pervaporation, used in the purification of the organic phase (compare Fig. 11), is an especially expensive membrane process for two reasons:

- the fluxes which can be achieved are rather low;
- the permeate has to be evaporated, i.e. the evaporation enthalpy has to be supplied and, afterwards, removed by condensation (Fig. 13).

Despite these disadvantages, the process is economical in this case, since the amount of water which has to be separated from the organic phase is relatively low. Fig. 14 shows, for one component of the organic phase regarded in these experiments as a component of a binary aqueous mixture, the high selectivity of the PVA-PAN composite membranes and the fluxes observed in these experiments. It should be noted that the flux of pervaporation is very sensitive to temperature, decreasing according to an Arrhenius-type law with decreasing temperature. Since the evaporation enthalpy is supplied by the liquid feed, a number of PV-modules and heat-exchangers have to be arranged in series in order to limit the temperature drop ΔT of the liquid in the module to reasonable figures of 5 - 10 K.

3.3. Treatment of RO-Concentrates

In the previous section, the potential of reverse osmosis in the separation of aqueous-organic solutions has been shown. There are, however, several limitations concerning the maximum achievable retentate concentration, mainly:

- the osmotic pressure of the retentate;
- the chemical stability of the membranes.

In addition, the permeate flux might become too low from an economical point of view with increasing concentration even before one of the other two limitations is reached.

With respect to osmotic pressure, a one-stage RO-process using the presently available modules is limited to a concentrate concentration of about 20 wt% organics. Depending on feed concentration, this figure can correspond to high recoveries, i.e. most of the feed-water might be separated. A retentate containing about 80 wt% water, however, can rarely be deposited or reused without further concentration. A further concentration can be achieved by:

- evaporation;
- distillation;
- combustion;
- pervaporation.

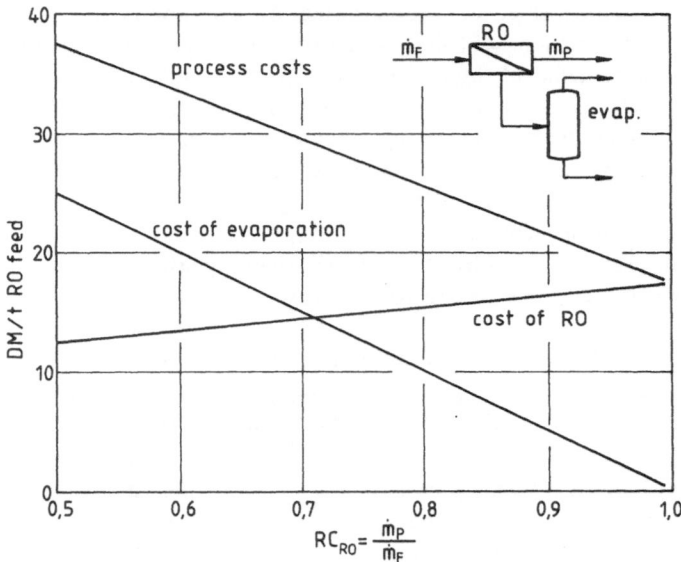

Fig. 15:   Influence of RO recovery on specific process costs.

Since all these processes are very expensive compared with RO, it can be concluded that, as a rule of thumb, the recovery rate of the RO stage should be as high as possible. This is demonstrated by Fig. 15, where the specific costs of a two-stage process, consisting of RO and evaporation, are plotted against the recovery rate of the RO stage. In this diagram, the specific costs of evaporation (cost per evaporation feed) are assumed to be 50 DM/m³, independent of capacity.

4. Conclusion

Composite RO membranes exhibit considerable potential in the separation of aqueous solutions of organics. In some cases, such as the treatment of landfill-drainage water, RO has already proved to be reliable and more economical than other alternatives.

Mass transport across the membranes cannot be described sufficiently accurately by the simple relationships used in sea- and brackish water desalination. Lower fluxes have to be expected.

In any case, the osmotic pressure will limit the retentate concentration to about 20 wt% organics. As a consequence, a further concentration, even in the case of disposal, has to be considered.

Chemical stability might also become a problem in high concentration ranges. Although the individual membrane materials are often resistant to very high concentrations of the organics, module parts like sealings or plates (in plate and frame systems) may fail. In order to solve this problem, modules specially suited for the treatment of aqueous-organic systems should be developed.

References

1 Rautenbach, R. and Albrecht, R., *Membrane Processes* (John Wiley) 1988.

2 Teipel, U., Studienarbeit, RWTH Aachen, Institut für Verfahrenstechnik, 1987.

3 NITTO-Firmenschrift R-3352-NBM, Sept. 1983.

4 Du Pont, Permasep. Products Engineering Manual (PEM), 1982.

5 Lloyd, D.R. and Dickson, J.M., ACS-Symp. Ser. 154, Washington D.C., 1981.

5 Toray Industries, Inc., Technical Bulletin.

7 Amsoneit, N., Technische Akademie Wuppertal, Jan. 20, 1985.

8 Rautenbach, R. and Herion, Ch., Aufarbeitung eines Deponie-Sickerwassers mittels Umkehrosmose, 1986.

9 Luning, J., Haus der Technik, Essen, March 2, 1985.

10 Slater, C.S., Ahlert, R.C. and Uchrin, G.C., Environ. Progress **2**, 4 (1983) 251.

11 Marquardt, K., Private communication.

12 Landratsamt Rastatt, Sickerwasserbehandlung, 1987.

13 Eisenmann Umwelttechnik, UT 12.

14 Rautenbach, R., Herion, Ch. and Janisch, I., Studie des IVT-RWTH Aachen und der Daimler Benz AG, Sindelfingen 1987.

15 Beyer, M., Diplomarbeit am IVT-RWTH Aachen, 1987.

Discussion

Chairman: H. Strathmann
Fraunhofer Institut, West-Germany

H. Strathmann • Thank you very much, Mr. Rautenbach, for this interesting paper, which shows that reverse osmosis application goes far beyond desalination of seawater or brackish water. It also shows how complex a system can be if you look at the real world of waste water treatment. The paper is open for discussion.

G. Schock • You mentioned that the cost of treating landfill drainage waste water was 13 DM/m³. What is the assumed lifetime of the membranes for these estimates?

R. Rautenbach • As far as I know, Hager & Elsässer assume a lifetime of two to three years.

G. Schock • You pointed out the good separation efficiency of composite membranes and you also mentioned the cleaning effect of sponge balls for tubular systems. Do you know of any combinations of both, i.e. tubular composite RO membranes with sponge ball cleaning?

R. Rautenbach • Up to now there are no units in operation with composite membranes and sponge ball cleaning.

G. Schock • Is there a reason for that?

R. Rautenbach • Well, the reason is simply that Wafilin only recently introduced tubular composite membranes.

H. Strathmann • You used reverse osmosis for the rejection of organic solvents. Organic solvents are usually present in very low concentrations in the waste water you mention. Wouldn't it be more economic to use a membrane which is selective for organic solvents (e.g. as in pervaporation), because the costs are proportional to the amount that goes through the membrane? All known RO membranes are selective for water. In pervaporation we also have solvent selective membranes. Do you think this would be possible in RO too or would you then use pervaporation?

R. Rautenbach • I would choose a process like pervaporation, yes. Membranes which are permeable to organics are available for pervaporation.

M. Campagna • 1) How do you see the future development of membranes themselves: has the field reached saturation or are new breakthroughs to be expected?   2) What is known about the long term performance of membranes?

R. Rautenbach • The first question actually addresses Heiner Strathmann. I personally don't think membrane development has reached a saturation level. But maybe Heiner Strathmann could tell us about further developments. — Your second question: the lifetime of the membranes, even in these waste water applications, can be expected to be several years.

In comparison with alternatives, the process pays off after about one and a half years. But in the more traditional fields of seawater desalination and boiler feed-water demineralisation there are some plants still operating with the first set of membranes after five or even ten years.

H. Strathmann • May I briefly comment on your question about membranes. A completely new type of membrane has recently been developed. The work was stimulated by pervaporation and gas separation. I think membrane development has a great future ahead of it.

Electrodialysis

W. McRae
Zürich, Switzerland

1. Background

The use of direct current electricity to increase the rate of dialysis of electrolytes or to produce demineralized water from potable water has been known for about 100 years. Early cells used three compartments between a single pair of electrodes, the compartments being separated from each other by (porous) diaphragms. The latter were essentially neutral and not intrinsically electrically conducting. Electrode reactions were important to this variety of electrodialysis which might better be regarded as a double electrometathesis, anions in the central electrodialysis compartment being replaced by hydroxide from the cathode compartment, cations by hydrogen ions from the anode compartment. It was noted that the choice of diaphragms affected the ultimate pH in the central compartment. Such three compartment ED cells were sold before World War II for water demineralization.

In Zürich in 1940, K.H. Meyer and W. Strauss[1] suggested a multiple compartment ED apparatus using many pairs of alternating non-porous, ion-selective membranes between a single pair of electrodes (Fig. 1). (The suggestion was based on studies by Prof. Meyer's group of the electricity generating organ in the electric eel). The group built a laboratory-scale proof-of-principle apparatus. The membranes were poorly ion-selective and unstable.

In 1948 at Harvard University, the author discovered how to make large structures of non-porous ion-exchange ("IX") resins and found that such structures were intrinsically electrolytically conducting, ion-selective and stable. The company Ionics, Inc. was formed later the same year, among other things to develop IX membranes. The discoveries were communicated by W. Juda (with whom the author worked at the time) to various IX resin scientists in the U.S. in 1948 and announced by Juda to the Gordon Research Conference on Ion Exchange in 1949. The latter announcement in particular stimulated research on synthesis by Permutit Ltd. (U.K.), Rohm and Haas Co. (U.S.), Permutit Co. (U.S.), the Dutch Government Research Organization T.N.O. among others. IX membranes in the apparatus of Meyer and Strauss made multi-compartment ED commercially feasible. (In modern usage "ED" means multi-compartment ED and three compartment cells in which electrode reactions are important to the process are now generally referred to as electrolysis cells).

In February 1952 Ionics demonstrated publicly the first small ED apparatus using IX membranes and in February 1953 the first commercial-scale apparatus. The first commercial ED apparatus was sold by Ionics in 1954 to Arabian-American Oil Co. ("ARAMCO") and installed in Saudi Arabia. In 1955 the Organization for European Economic Cooperation ("O.E.E.C.") funded development by the Dutch Organization for Applied Scientific Research ("T.N.O.") of ED process and apparatus (based on IX membranes). The development was licensed to Aqua-Chem in the U.S., to William Boby in the U.K., Bronswerk in Holland as well

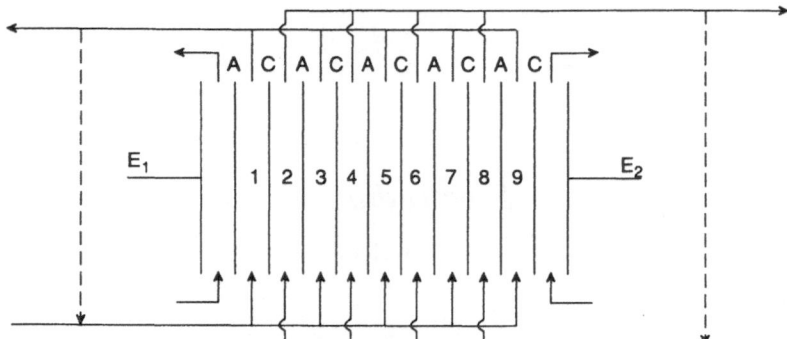

Fig. 1:   Schematic representation of ED/EDR
          A = Anion permeable, cation impermeable anion exchange ("AX")membrane
          C = Cation permeable, anion impermeable cation exchange ("CX") membrane
          Compartments 1, 3, 5, 7 and 9 are demineralizing when electrode E_1 is positive,
          concentrating when E_1 is negative.
          Compartments 2, 4, 6 and 8 are concentrating when E_1 is positive,
          demineralizing when E_1 is negative.

as to others. The T.N.O. ED group was disbanded in the 1950's, many of the participating scientists and engineers joining the ED group at the South African Council for Scientific and Industrial Research ("SACSIR"). As well as building in 1959 the largest ED plant at the time, the latter group summarized the state of the art in the monograph "Demineralization by Electrodialysis", ed. J.R. Wilson[2]. Almost three decades later it is still a very useful volume.

  Herein the following abbreviations are used:

- "ED": "Electrodialysis", a separation process using an apparatus consisting of a multiplicity of thin compartments between a single pair of electrodes. The compartments are defined by (non-porous) membranes permeable to small ions, every other membrane being substantially more permeable to ions of one charge sign than are the remaining membranes. The process is used to transfer low molecular weight ions from one liquid stream to another. Typical applications are demineralization of water and concentration of electrolytes. The energy for separation is supplied by direct electric current applied between the electrodes. It is not a filtration process; there is essentially no pressure gradient from one face of a membrane to the other;

- "EDR": "Reversing Type ED", an ED process in which the direct electric current is periodically (symmetrically or asymmetrically) reversed;

- "IX": "Ion Exchange", a process in which small ions in a first liquid solution are exchanged for other small ions of like charge sign from a solid or from a second liquid insoluble in the first liquid solution. The major uses are water softening and water demineralization;

- "AX": "Anion Exchange", an IX process in which negatively charged ions are exchanged;

- "CX": "Cation Exchange", an IX process in which positively charged ions are exchanged;

- "Stack": An ED or EDR module consisting of one or more pairs of electrodes, each such pair bounding a multiplicity of ion-permeable membranes, there being an electrolyte solution space between each pair of membranes;

- "RO": "Reverse Osmosis", "Hyperfiltration" or "Piezo-osmosis", a process for separating a low molecular weight solvent from a solution containing minor amounts of electrolytes and/or non-ionized solutes by forcing the solvent under pressure through a non-porous membrane which is permeable to the solvent but not substantially permeable to the solutes;

- "UF: "Ultrafiltration", a process for separating solvent and low molecular weight solutes (ionized or non-ionized) from colloidal and larger materials by forcing the solvent and low molecular weight solutes under pressure through a "nanoporous" membrane, i.e. a membrane having a mean pore size in the range of from ~ 1 to ~ 100 nanometers. The colloidal matter is generally not observable under a visible light microscope and not removable by ordinary centrifugation;

- "MF": "Microfiltration" or "Membrane Filtration", a process for separating solvents, solutes (ionized or non-ionized) and colloidal materials from suspended particulate matter by forcing the solvents, solutes and colloids under pressure through a "microporous" membrane, i.e. a membrane having a mean pore size in the range of from ~ 0.1 to ~ 1 micrometer. The particulate matter is generally observable under a visible light microscope and removable by ordinary centrifugation.

2. IX Membranes

2.1. Introduction

Commercial IX membranes are typically from 0.015 to 0.06 cm thick and available in areas from a fraction of a square meter to several square meters. Since the invention in 1948 of IX membranes by the author, a wide variety of such membranes has been brought into at least limited production by more than a dozen organizations, several of which have subsequently given up production. It appears that the types which have been commercially successful are those which are made by (or under contract for) a designer/manufacturer of ED equipment (e.g. Asahi Chemical Industry Co.; Asahi Glass Co; Ionics, Inc.; P.R. China; Tokuyama Soda Co.; U.S.S.R.) although some of the production may be sold to others. IX membranes (and ED stacks using them) are not commodities, unlike RO and UF modules. Further, ED stacks tend to be systems, i.e. the membranes and mass-transfer-promoting intermembrane separators are designed around each other. Even if they are of the same size, the membranes and/or separators of one manufacturer are generally not interchangeable with those of another.

In the following paragraphs the various types of commercially available, hydrocarbon based membranes are discussed briefly. No attempt is made to give recipes for manufacture. Perfluorinated IX membranes are not discussed. The classification system used is that convenient to the author. Others might choose a different, but equally useful system.

2.2. Fabric-Reinforced, Isoporous Gel, Homogeneous Membranes

These were the original IX membranes invented by the author in 1948. The first such membranes were based on polycondensation products, e.g. polymers of phenolsulfonic acid, phenol and formaldehyde but were soon replaced by membranes based on addition polymers (e.g. vinyl polymers) following the teachings of the patents of the late J.T. Clarke of Ionics[3]. For example, a mixture of styrene, divinylbenzene (the commercial material containing substantial amounts of ethylvinylbenzene and diethylbenzene), toluene (as an in-situ swelling agent for the eventual polymer) and benzoyl peroxide (as polymerization catalyst) is cast over a suitable fabric on a glass plate, covered with a second glass plate and the resulting sandwich heated at polymerization temperature until completely polymerized. The glass plate sandwich serves to prevent substantial loss of the swelling agent from the polymerizing sheet, such conservation being essential. The thickness of the reinforcing fabric determines the thickness of the membrane. After polymerization, the swollen, fabric-reinforced polymer sheet is recovered, the aromatic swelling agent replaced with a solvent more suitable for subsequent steps in the manufacture. After this the sheet is, for example, sulfonated to make cation exchange ("CX") membranes or chloromethylated and trimethylaminated to make anion exchange ("AX") membranes. The principal variables are the thickness of the reinforcing fabric, the amount of crosslinking agent (e.g. divinylbenzene) and the quantity of in-situ swelling agent. These affect the electrical resistance per unit area and the diffusion and electro-osmotic transfer rates of medium molecular weight, non-ionized solutes and solvents in the solutions electrodialyzed.

The IX resins of these membranes are not homogeneous on a microscopic scale, the water content and IX capacity varying over distances of ~ 0.01 µm, but they are homogeneous on a macroscopic scale and probably less heterogeneous on a microscopic scale than other IX membranes. Neither are they strictly isoporous, but again more so than other IX membranes (see below). Pore sizes can be varied over a range which, on one end, cuts off the diffusion of molecules of the size of sucrose and, on the other end, permits almost unhindered diffusion of such non-ionized molecules.

The fabric-reinforced, gel IX membranes tend to be brittle and to have a high electro-osmotic water transfer but are comparatively inexpensive, long-lived and commercially successful in the ED of brackish and potable water and of concentrated cheese whey. It is estimated that there are 850,000 to 900,000 m² of such membranes installed.

Figs. 2 and 3 are schematic representations of suitable crosslinked CX and AX polymers respectively. The polymers are probably more or less random, entangled and interpenetrated networks. There are probably no pores in the sense of capillaries but rather water-filled, interconnected cavities among the polymer segments, the sizes of the cavities clustered around some mean determined by the parameters of synthesis of the membranes. The fixed charged groups (sulfonate in Fig. 2, quaternary ammonium in Fig. 3) are not free to migrate. Mobile counter-ions (M^+ and X^- resp.) carry the electric current through the membranes. Mobile ions of the same charge sign (co-ions) as the fixed charges do not readily permeate the membranes, the extent of permeation correlating reasonably well with the modified Donnan Exclusion treatment of E. Glueckauf[4]. The mobilities of the various counter-ions in the membranes are substantially less than in aqueous solutions.

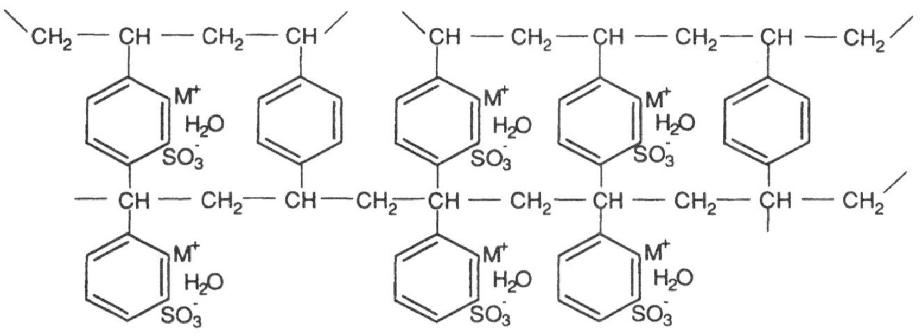

Fig. 2: Schematic representation of cation exchange resin.

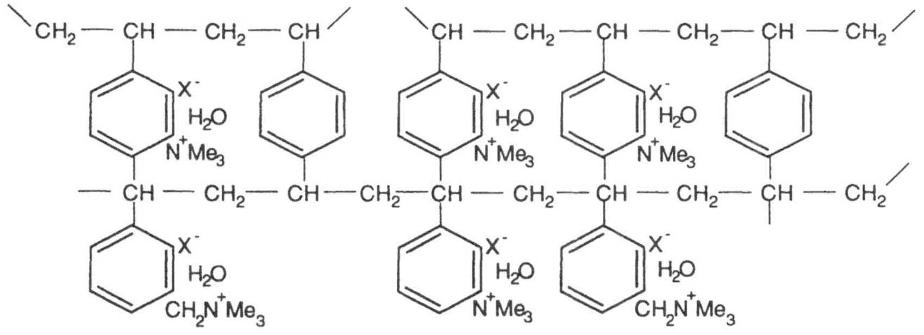

Fig. 3: Schematic representation of anion exchange resin.

2.3. Fabric-Reinforced, Bonded, (Macro)heterogeneous Membranes

These membranes are prepared by blending ~ 75 to 80 parts of pre-formed IX resin powder, 20 to 25 parts of a thermoplastic film-forming material (such as polyethylene, PVC, vinylchloride-vinylacetate copolymers, butadiene-styrene copolymers) with or without solvents and/or plasticizers on conventional roll mills or in conventional kneaders and then calendering on suitable fabrics. The IX resin powder may be prepared by grinding commercial IX beads or by using submicrometer beads. In the early IX membranes of this type, the IX powders were not much less than ~ 100 μm in average diameter and the membranes suffered from loss of IX resin from the surface, internal separation of binder and IX particles allowing pockets of water internally and decreased ion-selectivity. There was also a problem with low limiting current density (see below). Modern heterogeneous IX membranes suffer less from these problems, apparently owing to more careful attention to selection of IX resins, particle size (1 to 10 μm), mechanical properties of the binder and adhesion between binder and IX resin. Modern heterogeneous IX membranes appear homogeneous to the unaided eye and have been successfully used in ED apparatus in China and Russia on a large scale. There are probably well in excess of 1 million square meters of such membranes installed.

2.4. Macrohomogeneous-Microheterogeneous Membranes

These are generally fabric reinforced, though unreinforced materials are available from some manufacturers. The category includes many subspecies comprising interpolymer mixtures, interpenetrating networks, block copolymers, snake-in-cage mixtures and the like, heterogeneous on a scale of 0.001 to 0.1 µm but homogeneous on a large scale. They are frequently opaque at visible wavelengths.

One group is made, for example, by dissolving linear polystyrene or butadiene-styrene copolymers in mixtures of divinylbenzene, styrene and dimethylphthalate, calendering onto reinforcing fabrics, polymerizing and then sulfonating or chloromethylating and trimethylaminating. Alternatively the mixture may be polymerized in a block, skived into (unreinforced) sheets and then activated.

Membranes have been made from block copolymers of, for example, styrene and butadiene or isoprene into which sulfonic acid or quaternary ammonium groups have been introduced by the usual methods. The swelling by water of the polystyrenesulfonic acid or polyvinylbenzyl-trimethylammonium blocks is restrained by the crosslinking due to crystallites of, for example, polybutadiene.

The combined installed membrane area of the above two subspecies is probably in the range of 700,000 to 800,000 square meters.

Other block copolymer membranes have been made from films of low density polyethylene by treating with, for example, sulfur dioxide and chlorine, subsequently hydrolyzing to sulfonic acid or reacting with dimethylaminopropylamine and then methyl bromide. The reaction conditions limit the ionic groups to the amorphous regions of the polymer. The crystalline regions provide the resistance to swelling by ambient polar liquids.

Interpenetrating network membranes have been made by swelling low or high density polyethylene in a mixture of divinylbenzene and styrene, polymerizing with benzoyl peroxide and heat or with gamma radiation and subsequently sulfonating or chloromethylating and trimethylaminating in conventional ways. The styrene-divinylbenzene copolymers (and therefore the ionic groups) interpenetrate the amorphous regions of the polyethylene. It has been claimed that the styrene-divinylbenzene copolymers are grafted on to the polystyrene, but it has been shown that no grafting occurs when polyethylene is replaced with polyvinylidene fluoride, and yet the aging characteristics are similar. Further, it has been shown that only a very small amount of, or even no divinylbenzene is required, the polystyrene polyelectrolyte apparently being entrapped in a snake cage. These membranes are not generally reinforced.

Interpolymer membranes have also been made by casting a mixture of about 1 part of a linear polyelectrolyte (e.g. polystyrenesulfonic acid or polyvinylbenzyltrimethylammonium chloride) and ~ 3 to 5 parts of a film-forming polymer (e.g. a copolymer of 40 parts acrylonitrile and 60 parts vinylchloride) from a mixed solvent (e.g. cyclohexanone and methanol). Both reinforced and unreinforced membranes of this type were made commercially for a few years.

Other interpenetrating network membranes have been made by plasticizing polyvinyl-chloride with dioctyl phthalate, styrene (or vinylpyridine) and divinylbenzene, coating on a suitable fabric, polymerizing and then sulfonating or quaternizing in the usual ways.

There are probably 200,000 to 250,000 square meters of installed membrane area of this type.

3. Some Ion Exchange Membrane Suppliers

1. Alma-Ata Electrochemical Works
    Ul. Zemnukhova 9A
    480028 Alma-Ata
    Kazakh S.S.R.

2. Asahi Chemical Industry Co., Ltd.
    Hibiya Mitsui Bldg.
    1-2 Yuraku-cho 1-chome
    Chiyoda-ku, Tokyo 100, Japan

3. Asahi Glass Co., Ltd.
    1-2, Marunouchi 2-chome
    Chiyoda-ku, Tokyo 100, Japan
    (The above two companies are not related)

4. Institute for Applied Research
    Ben Gurion University of the Negev
    P.O. Box 1025
    Beersheva, Israel 84110

5. IONAC Chemical Div.
    SYBRON Corp.
    Birmingham, NJ 08011, USA

6. Ionics, Inc.
    65 Grove St.
    Watertown, MA 02172, USA
    (The above two companies are also not related)

7. RAI Research Corp.
    225 Marcus Blvd.
    Hauppauge, L.I., NY 11788, USA

8. Second Institute of Oceanography
    National Bureau of Oceanography
    P.O. Box 75
    Hangzhou, P.R. China

9. Tokuyama Soda Co., Ltd.
    No. 1, 1-chome, Mikage-cho
    Tokuyama City, Yamaguchi 745
    Japan

4. ED Apparatus

The heart of ED apparatus is the ED stack consisting of one or more pairs of electrodes and, between each pair, many (100 to 1000) ion-permeable membranes ranging in size from about 0.01 (for laboratory apparatus) to 1 or more square meters for commercial apparatus. The membranes are typically spaced from ~ 0.035 to 0.2 cm apart by gaskets forming solution-containing chambers between the membranes. Typically the membranes are alternately AX (therefore anion-selective) and CX (therefore cation-selective) (compare Fig. 1) although infrequently one of the alternating membranes (usually the AX) may be replaced by an essentially neutral (i.e. non-selective), hydraulically impermeable membrane. In either case, upon passage of a direct electric current between the electrodes in either direction, every other chamber (compartment, cell, space) will be partially deionized (demineralized, desalted), the intervening compartments receiving the electrolyte removed from the deionizing compartments. The latter are typically called diluting compartments, and the electrolyte receiving compartments are called concentrating compartments. The compartments have inlet and outlet means (manifolded internally within the stack) permitting solutions to flow through them. The flow through all the diluting compartments between the pair of electrodes may be in parallel or the compartments may be divided into 2 or more blocks ("stages"), solution flowing first in parallel through the diluting compartments of one block, recombining and then in parallel through the compartments of the second block etc.

Since the membranes are generally ion-selective, 90 to 95% of the electric current is usually carried through the AX membranes by (negatively-charged) anions (such as chloride, sulfate, bicarbonate, nitrate, fluoride) and through the CX membranes by (positively-charged) cations (such as sodium, calcium, magnesium, ammonium). In the solutions flowing through the chambers, usually only 50 to 60% of the electric current is carried by anions and the remainder

(40 to 50%) by cations. As a result, the interfaces between the IX membranes and the solutions in the diluting compartments become more or less strongly depleted in electrolyte, whereas the interfaces between the membranes and the solutions in the concentrating compartments become more or less highly concentrated (see Fig. 4). At the latter interfaces, the solubility limits of sparingly soluble salts (such as calcium carbonate and calcium sulfate) can be exceeded and scales of such salts formed on or in the membranes. At the interfaces in the diluting compartments, the concentration of electrolyte can approach zero as the electric current is increased. If the current is increased still further, it will be carried in part by ions resulting from the dissociation of water, i.e. hydroxide ions (OH⁻) through the AX membranes,

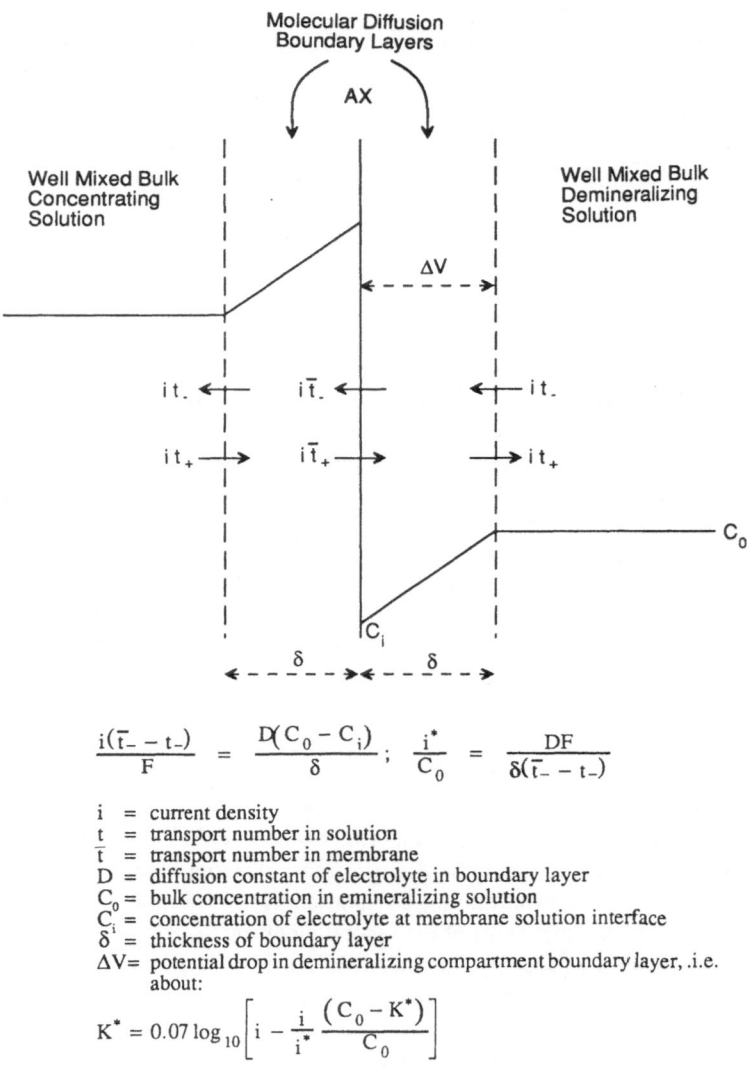

$$\frac{i(\bar{t}_- - t_-)}{F} = \frac{D(C_0 - C_i)}{\delta} \; ; \; \frac{i^*}{C_0} = \frac{DF}{\delta(\bar{t}_- - t_-)}$$

i = current density
t = transport number in solution
$\bar{t}$ = transport number in membrane
D = diffusion constant of electrolyte in boundary layer
C_0 = bulk concentration in emineralizing solution
C_i = concentration of electrolyte at membrane solution interface
δ^i = thickness of boundary layer
ΔV= potential drop in demineralizing compartment boundary layer, .i.e. about:

$$K^* = 0.07 \log_{10}\left[i - \frac{i}{i^*} \frac{(C_0 - K^*)}{C_0} \right]$$

Fig. 4:   Transport mechanism in electrodialysis.

hydrogen ions (H$^+$) through the CX membranes. The former can aggravate scaling by calcium carbonate. Nevertheless, in the case of demineralization of potable and brackish waters, it is desirable to operate at high current densities (i.e. high electrolyte transfer per unit area) in order to reduce investment costs. Mixing the solutions at the membrane interfaces with bulk solution in the interior of the compartments permits higher current densities to be used without substantial dissociation of water or scaling by poorly soluble salts. In principle, such mixing can be achieved by flowing the solutions through the compartments at rates which are in the naturally turbulent range (i.e. high Reynolds numbers), and this has in fact been done on a pilot scale. All commercial apparatus, however, use some kind of mixing promoting structure (turbulence promoters) in the compartments. The limiting current (i.e. the maximum current which is substantially free of water dissociation) is found to be essentially directly proportional to the concentration in the diluting compartments for any given electrolyte or mixture of electrolytes. Even with turbulence promoters it has been found to be difficult to achieve limiting currents of more than about 500 amp. per square meter, for e.g., a 0.05 normal sodium chloride solution at room temperature, (this corresponds to an unstirred film for molecular diffusion at the membranes averaging perhaps 0.004 cm).

It is also desirable to use small separators between membranes and thin membranes (at least in the case of concentrated solutions) to reduce the electrical resistance of an ED stack and therefore the energy consumption to remove a given amount of electrolyte. This has led ED designers to use structures, which can both support the thin membranes and promote mass transfer, as separators (spacers) between membranes. At present there are only two general types of such spacers in use:

- The tortuous or labyrinth path used by Ionics, Inc. and some designers in the U.S.S.R. These have one or more channels (typically ~ 10 mm wide) winding back and forth across the surfaces of the membranes. Mixing promoting obstructions typically of half the thickness of the spacer are positioned about every centimeter or so. To achieve high limiting currents, flow rates of ~ 20 to 40 cm^3/sec per cm^2 of flow path cross-section are used. Flow path lengths of a few hundred centimeters (e.g. 350 to 500 cm) permit typically 40 to 50% removal of electrolyte in a single pass through the spacer. Such spacers are inexpensive to make, generally being assembled from plies stamped from thermoplastic sheet, e.g. low density polyethylene or polymeric plasticized polyvinylchloride. They are quite forgiving of particulate matter in the solutions processed.

- The sheet flow spacer is used by all designers other than Ionics and the above mentioned designers in the U.S.S.R. These have a marginal ("picture frame") gasket area, the central open area containing a woven or non-woven plastic screen or expanded plastic sheet. Flow enters at one corner or edge and leaves at an opposite corner or edge. Care must be taken in the design to have essentially pistonlike flow across the surfaces of the membranes, i.e. no areas which have velocities substantially lower than average. These spacers tend to be more expensive than the tortuous path spacers (in some cases much more expensive), but give better support to thin membranes. At the same limiting current densities, flow rates and pressure losses are generally less than is the case with tortuous path spacers. Usually 40 to 50% removal of electrolyte can be achieved in a single pass through the spacer. Asahi Glass Co. and Tokuyama Soda Co. have, however, made screen type spacers capable of 90 to 99% removal in a "single pass". It is likely that these use folded path sheet-flow separators as well as internal flow equalizing devices.

Even with such turbulence promoting spacers, it was often found to be desirable to acidify the solution in the concentrating compartments (to control calcium carbonate) and/or to add threshold precipitation inhibitors such as sodium hexametaphosphate (to control calcium sulfate). At any rate, it was usually necessary to acidify the solution contacting the negatively charged electrode (cathode) to prevent precipitation of calcium carbonate and magnesium hydroxide. In 1956 the writer invented reversing type ED ("EDR") based on the fact that, in an ED stack with symmetric membranes and equal thicknesses of diluting and concentrating compartments, the choice of direction of direct current is arbitrary (compare Fig. 1). Those compartments which are demineralizing with a current in one direction become concentrating when the current is reversed. Such reversal has generally eliminated the need for continuous feed of chemicals to any of the compartments (including the cathode compartments) of ED stacks even when the solution in the concentrating compartments is supersaturated in calcium sulfate and calcium carbonate. Many designers now utilize this concept or offer it as an option. Japanese designers seem to have been slow in using this concept. Ionics, Inc. does not use EDR in cheese-whey deashing probably because of the loss of whey substance during the reversal process. For similar reasons, EDR is not used in seawater concentration apparatus.

With respect to electrodes for ED, most designers use as anodes titanium sheet (infrequently other valve metals) thermally or electrically plated with platinum. Cathodes in non-reversing ED are generally stainless steel or other corrosion resistant alloys such as Hastelloy C. The latter are not suitable for EDR and are replaced with platinum-plated titanium or similar electrodes. Designers in the P.R. China use small, vertical rods (~ 0.3 cm diameter) thermally plated with ruthenium as anodes in non-reversing ED and for both electrodes in EDR. Graphite is still used by some designers (though infrequently) for both electrodes in EDR. Even though graphite is consumed as an anode, it appears to be competitive in cost with noble metal coated titanium in EDR use.

5. Some Suppliers of ED Apparatus

1. Alma-Ata Electrochemical Works
    (see address in section 3)

2. Asahi Chemical Industry Co., Ltd.
    (see address in section 3)

3. Asahi Glass Co., Ltd.
    (see address in section 3)

4. Berghof GmbH
    Harrelstrasse 1
    D-7412 Eningen u.A., B.R.D.
    (The designs of Berghof, G.K.S.S. and
    Stantech (see below) are quite similar)

5. Commissariat à l'Energie Atomique
    (C.E.A.)
    B.P. No. 1
    13115 St Paul les Durance, France

6. G.K.S.S. Forschungszentrum
    Geesthacht GmbH
    Max-Planck Strasse
    D-2054 Geesthacht, B.R.D.

7. HPD Inc.
    Div. CBI Industries
    1717 North Napier Blvd.
    Naperville IL 60566, U.S.A.

(HPD Inc. is a licensee of Asahi Glass Co.)

8. Ionics. Inc.
    (see address in in section 3)

9. Portals Water Treatment Ltd.
    Permutit House
    632/652 London Rd.
    Isleworth Middlesex
    England TW7 4EZ

10. Second Institute of Oceanography
    (see address in section 3)

11. Société de Recherches Techniques et
    Industrielles (S.R.T.I.)
    B.P. 10
    78530 Buc, France

12. Stantech, Inc.
    Div. Membrane Technology and
    Research, Inc.
    1360 Willow Rd.
    Menlo Park CA 94025, U.S.A.

13. Tokuyama Soda Co., Ltd.
    (see address in section 3)

6. Applications of ED

There are two main applications of ED.

6.1. Demineralization of Potable and Brackish Water

There are some 4000 plants (of both reversing and non-reversing types) divided very roughly as follows:

- P.R. China, about 2000 plants, more than 600,000 m^3/d, more than 1 million m^2 of installed membrane area;
- Ionics, Inc., about 1500 plants, more than 450,000 m^3/d, more than 900,000 m^2 of installed membrane area;
- U.S.S.R., about 500 plants, more than 150,000 m^3/d, more than 300,000 m^2 of installed membrane area;
- Miscellaneous (including Asahi Chemical Industry Co.; Asahi Glass Co.; Portals Water Treatment; S.R.T.I.; Tokuyama Soda Co.; Ltd.), 20 to 40 plants, 18,000 to 24,000 m^3/d, 36,000 to 48,000 m^2 installed membrane area.

One segment of this application is the partial demineralization (e.g. about 90% salt removal) of essentially potable water as pre-treatment before mixed-bed IX de-ionizers for pure and ultrapure water applications, e.g. rinsing of electronic components, pharmaceutical applications, feed to high pressure boilers. Another segment is the removal and concentration of nitrate from potable water.

6.2. Concentration of Seawater

Concentration of seawater to 18 to 20% for production of salt by evaporation and crystallization is carried out in about 10 generally very large plants producing a total of about 1,600,000 metric tons of salt per year and involving about 800,000 m^2 of membrane area. Seven plants are in Japan, one each in S. Korea, Taiwan and Kuwait. None are justified on economic grounds compared to imported solar or mined salt. (It is Japan government policy that all food-grade salt, in contrast with industrial salt, must be made in Japan and sold to the Japan Salt Monopoly Corporation). The membranes are laminar or coated to make them selective to univalent ions. Reversal is not used. The plants are divided roughly as follows:

- Asahi Chemical Industry Co., about 5 plants, about 800,000 metric tons salt per year, about 400,000 m^2 membrane area;
- Asahi Glass Co., about 2 plants, about 370,000 metric tons salt per year, about 185,000 m^2 membrane area;
- Tokuyama Soda Co., about 3 plants, about 400,000 metric tons salt per year, about 200,000 m^2 membrane area.

An application, closely related to the above, is the simultaneous concentration of industrial effluents and recycle of recovered water, developed as a result of the trend toward zero industrial discharge. Examples are the concentration of blow-down from cooling towers in power plants and the processing of metal treatment wastes. Note that the osmotic pressure difference between 19% sodium chloride brine (as recovered in the electrodialytic concentration of seawater) is ~ 200 atm. at 25 °C, well beyond the practical range for reverse osmosis.

6.3. Miscellaneous Applications:

- *Desalting of cheese whey and non-fat milk.* There are perhaps 65 to 70 such ED plants of which Ionics has ~ 40, S.R.T.I. ~ 20, most of the remainder divided among the three above mentioned Japanese companies. Total installed membrane area is perhaps 35,000 square meters.

- *Recovery of valuable components from metal plating or treating effluents.* There are as yet very few such ED plants, although it is often economic to recover valuable metal salts in this way instead of purchasing new salts.

- *De-ashing of beet, cane or other sugar juices and molasses.* AX membranes are generally subject to fouling by medium molecular weight organic carboxylic acids in such solutions. Fouling resistant AX membranes have been proposed, but the short processing season for such sugar solutions leads to high capital charges. Further, the additional sugar which could be recovered through desalting is not justified in view of the agricultural policies of the sugar producing countries. There are few, if any, ED plants desalting sugar solutions.

- *Demineralization of industrial and municipal primary, secondary and tertiary sewage effluents.* There have been a few pilot plants for this application, but few, if any, ED plants on a commercial scale yet.

- *Deacidification/acidification of fruit juices.* There are few, if any, such plants on a commercial scale, although there have been pilot applications.

- *Desalting of soy sauce, amino acid solutions, fermentation products etc.* There have been pilot applications and there are probably a few small ED plants for such applications.

- *Demineralization of blood plasma.* There are few such plants.

- *Demineralization of seawater.* There are a few small ED plants for this application.

6.4. Piezodialysis ("PD")

This process, which is still in an early research stage, uses micromosaic IX membranes (i.e. membranes having contiguous regions of AX and CX resin, each region penetrating from one surface of the effective membrane to the other, and each region having a characteristic diameter of 1 mm or less). The driving force is a pressure difference as in the case of RO, but the permeate is more concentrated in electrolyte than the retentate (similar to ED). In the case of desalting of potable and brackish waters, the permeate can be only a small fraction of the retentate and, in principle, the irreversible energy requirements can be lower than either ED or RO. Further, the demineralized product is the retentate and remains under pressure. Very little R & D has been expended on the process compared to the sums which have been spent on ED and RO. It is believed that Toyo Soda Manufacturing Co. Ltd. may have a feasible membrane.

6.5. Bipolar, "Water-Splitting" Membranes

Such IX membranes were apparently first disclosed by J. Frilette[5], and by E. Glueckauf and G.P. Kitt[6]. There has been a major effort by Allied-Signal Corp. for about the last 15 years (involving many millions of dollars) to develop feasible bipolar membranes and applications. More or less conventional ED apparatus may be used. Bipolar membranes consist of AX resin on one surface and CX resin on the other (see e.g. Refs. 7 and 8). When a direct current is passed in the direction which pulls anions away from the AX/CX interface through the AX resin, then the interface between the AX and CX regions will become essentially completely polarized,

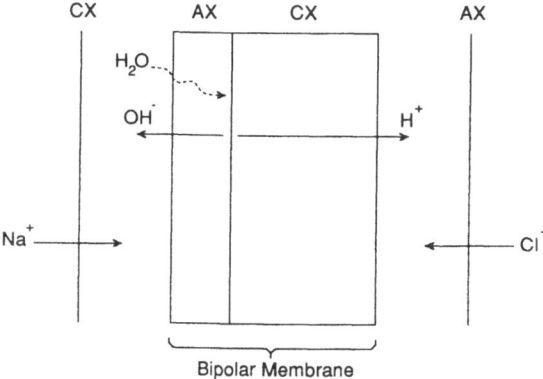

Fig. 5:   Schematic functionality of bipolar membrane.

i.e. depleted of electrolyte (see Fig. 5). If nevertheless direct current continues to be passed it can only be carried by OH$^-$ and H$^+$ ions resulting from the spontaneous and induced dissociation of water (plus, of course, leakage of cations through the AX resin and of anions through the CX resin). In principle the AX resin in a bipolar membrane emits OH$^-$ and the CX resin H$^+$. Compared with electrolysis of water, no H$_2$ and O$_2$ gases are produced. The reversible electrical potential is ~ 0.8 volts (assuming both OH$^-$ and H$^+$ are produced by the apparatus at one gram-equivalent per liter), ~ 1.23 volts less than the reversible potential for an electrolysis producing H$_2$, O$_2$ as well as H$^+$ and OH$^-$. (The usual H$_2$/O$_2$ electrolytic generator of course produces no net H$^+$ or OH$^-$). In principle, the materials of construction for a water splitting ED cell at the same current density are much less expensive than for a (three compartment) water electrolysis cell. However, the specific electrical resistance of the essentially completely demineralized AX/CX interface must be quite high (though very thin, perhaps 1 to 10 nm). It has been shown that the spontaneous ability of water at such polarized interfaces to dissociate into H$^+$ and OH$^-$ in the absence of the Second Wien effect is very much less than the H$^+$/OH$^-$ fluxes actually generated. The dissociation of water can be catalyzed by primary, secondary and tertiary amines and/or carboxylate groups. Under the high fields prevailing at the interface between the AX and CX regions (perhaps 1 to 10 million volts/cm over distances of perhaps 1 to 10 nm)[10,11], quaternary ammonium groups are apparently degraded to tertiary amine groups which catalyze water dissociation. There may also be a tendency under such high fields for segments of AX and CX polymer to migrate into the respective oppositely charged regions, resulting in internal salts and partial dehydration of the interface, (tertiary amine groups in the base form are also less hydrophilic than quaternary ammonium groups). As a result of these processes in the very aggressive environment at the AX/CX interface, the electrical resistance of present bipolar membranes increases at such a rate that the economic life of the membranes is typically only about a year. It is probable that the last word is not yet in on the synthetic (or physical) chemistry of the interfaces of bipolar membranes.

Other water splitting ED systems have been studied. For example, if demineralizing compartments of an ED apparatus (consisting of repeating units of three or more compartments) are filled with polyelectrolyte having a size larger than the pores of the IX membranes or with particulate IX resin, then the interface at one of the membranes (depending upon the charge of the polyelectrolyte or particulate IX resin) will also be water splitting. The Japanese univalent ion-selective membranes (for seawater concentration) are bipolar membranes and will behave as such at high current densities in dilute solutions. At least some of the Japanese antifouling membranes are also bipolar. AX membranes which have been fouled in use by organics are bipolar and water splitting (see e.g. Ref. 12). Deliberately fouled IX membranes (e.g. to render them selective to univalent ions or antifouling against other foulants) are bipolar and water splitting.

Should it, therefore, be desirable to split water into OH^- and H^+ without simultaneous production of H_2 and O_2, several ED options are available.

7. Some Design and Operating Considerations in ED

IX membranes are effective in ED because they carry electric current almost exclusively by means of ions of one charge sign or the other, CX membranes by means of cations, AX membranes by means of anions. In contact with solutions having concentrations of ~ 0.1 N or less, typically $\sim 95\%$ of the current is carried by such (counter) ions. This contrasts with the situation in ambient electrolyte solutions, (e.g. in sodium chloride solutions), $\sim 40\%$ of the current is carried by sodium and 60% by chloride. As shown in Fig. 4, this means that upon passage of one gram-equivalent of electric current there will be a deficiency at AX membranes of $0.95 - 0.60 = 0.35$ gram-equivalents of chloride and at CX membranes of $0.95 - 0.40 = 0.55$ gram-equivalents of sodium. These deficiencies can be supplied by diffusion from the bulk of the solution through some molecular diffusion layer ("δ") in which eddy-diffusion is small compared to molecular diffusion. The maximum possible rate of molecular diffusion can occur when the concentration at the interface between the membrane and the solution approaches zero. As shown in Fig. 4, the ratio of the current density ("i") to the bulk concentration ("C_o") should only be a function of: Faraday's Constant ("F", ~ 26.8 Ah, the quantity of electric current equal to one gram-equivalent); the diffusion constant ("D") of the salt; the difference ("$\bar{t} - t$") between the fraction of current carried by the exchangeable ion in the membrane and in solution; and the molecular diffusion thickness parameter ("δ"). (In the case of simple geometries ("δ") can be estimated from heat or mass transfer data in the literature. For more complicated geometries it is really an adjustable parameter which fits the data to the theory. The model of Fig. 4 does not contain a term for convection of ions by electro-osmotic water transfer through the membrane with the counter ions. It can be shown that, at least in the case of sodium chloride solutions and commercial AX membranes, correction for such convection is not significant at bulk concentrations below ~ 0.25 N, i.e. $\sim 15,000$ ppm).

In accordance with the simple model of Fig. 4, the value of the current density "i" when the interfacial concentration approaches zero is defined as the limiting current, here written "i^*". As i^* is approached, a substantial increase in the electrical resistance of the apparatus should be experienced (the actual increase in resistance should be corrected for any increase due to demineralization in the apparatus). An estimate of the expected resistance increase is shown in Fig. 4 in terms of ΔV, the potential drop in the demineralizing compartment boundary layer. "K^*" represents an estimate of the residual specific conductivity of the interface at polarization. Depending upon the actual value of the bulk concentration "C_o" and how "i^*" is actually measured, the maximum value of ΔV at i^* is a few tenths of a volt. The above (simplified) theory predicts that anion-selective membranes should reach i* at current densities ~ 50 to 60% greater than for cation selective membranes. In addition to sodium chloride ions, hydrogen and hydroxide ions from dissociation of water are present at the solution-membrane interfaces. The latter ions should carry an increasing fraction of the current as the concentrations of sodium and chloride ions at the interfaces approach zero. Hydrogen ions pass through a CX membrane and hydroxide ions through an AX Membrane (as their respective i*'s are approached). This has been shown experimentally. In the case of known commercial CX membranes, a flux of hydrogen ions can be measured as i approaches and exceeds i* but it appears to reach a limiting flux which is not of great practical significance to the ED process. As the current at a cation-selective membrane increases beyond i*, most of the incremental current appears to be carried by sodium ions. However, for commercial AX membranes, as i* (for the AX membrane) is approached and exceeded, a considerable fraction of the incremental current above i* is carried by hydroxide. Simons[13] attributes this to catalysis of water dissociation by weakly basic amines present or formed in the surface of the AX membranes by localized high fields (1 to 10 million volts/cm over distances of some nm) at currents near and above i*. The current density corresponding to the onset of significant hydroxide ion flux is known as the polarizing current density for the apparatus. It is seldom exceeded in ED of brackish water because of the frequent occurrence in the solutions treated of substances (e.g. calcium bicarbonate) which can precipitate on the AX membranes at pH's different from 7.

For ED of dilute solutions, a better economic balance between investment and operating cost might be possible if AX membranes were available which transferred very little hydroxide from the dissociation of water at currents in excess of i*, (i* can often be exceeded without significant penalty in the ED of potable water). Referring to Fig. 4, the polarizing (limiting) current density at an AX membrane is an inverse function of δ. Designers therefore try to minimize δ. In principle, naturally turbulent flow could be used. However, all commercial ED stacks use membranes spaced apart by from ~ 0.3 to ~ 3 mm and separated by turbulence promoting structures. These are essentially passive mixers at intervals of from ~ 1 to ~ 10 mm. Those with the larger intervals tend to be proprietary structures die-cut from thermoplastic sheet. Those with the smaller intervals tend to be commercially available woven plastic screens or expanded plastic sheets. In either case, the maximum practical value of i* (expressed in A/cm^2) is generally found to be numerically equal to C_o (the bulk concentration in the diluting compartments, expressed in gram-equivalents per liter). For example, if C_o ~ 0.05 N then i* will be ~ 0.05 A/cm^2. In the case of dilute solutions (less than ~ 0.1 N), the principal electrical resistance will be in the stream being demineralized. If the distance between the membranes is about 1 mm, then, since the equivalent ionic conductance of sodium chloride solutions is ~100, the limiting current in such dilute solutions corresponds to a potential drop of ~ 1 volt per cell-pair (defined as an AX membrane, a demineralizing space, a CX membrane and a concentrating space). The current efficiency for electrolyte transfer is roughly 90%. From the above it may be calculated that at ~ 1 volt per cell-pair, ~ 0.5 kWh are required to remove 1 kg of sodium chloride from a dilute skeleton. One volt per cell-pair and 0.5 kWh/kg of salt are very useful rules-of-thumb. For example to demineralize brackish water from ~ 1500 to ~ 500 ppm one might expect to require ~ 0.5 kWh/m^3 of product. The 0.5 kWh rule may be easily adjusted for other equivalent weights. When the solution demineralized is ~ 0.1 N, only about half the electrical resistance of the membrane stack will be found in the demineralizing solution. One could then use ~ 2 volts per cell-pair before reaching i*. In practice, however, an economic balance between investment and operating and maintenance costs indicates that 1 volt and therefore 0.5 kWh are still good rules-of-thumb. At seawater concentrations, an economic balance favours ~ 0.25 kWh/kg of salt removed.

The above rules-of-thumb refer to electrical energy directly used for transferring electrolyte. There will be in addition energy used for auxiliaries, e.g. pumping. This is usually ~ 0.5 kWh/m^3, also as a rule-of-thumb. If reversing type ED is used there will, in general, be no requirement for continuous feed of chemicals to control scaling and fouling (though with some solutions the apparatus may have to be chemically cleaned occasionally).

Electrodialysis with ion-exchange membranes is a long established, highly successful process. It is particularly economic compared to other separation processes for demineralizing dilute solutions and/or producing very concentrated solutions of electrolytes.

References

1   Meyer, K.H. and Strauss, W., Helv. Chim. Acta **23** (1940) 795-800.

2   J. Wilson, ed., *Demineralization by Electrodialysis* (Butterworths Scientific Publ., London) 1960.

3   Clarke, J.T., U.S. Pats. 2,730,768; 2,780,604; 2,800,445.

4   Glueckauf, E., Proc. Royal Soc. **A 268** (1962) 339 and 350.

5   Frilette, J., J. Phys. Chem. **60** (1956) 435.

6   Glueckauf, E. and Kitt, G.P., J. Appl. Chem. **6** (1956) 511.

7   Leitz, F.B., U.S. Pats. 3,562,139 and 3,654,125.

8   Dege, G.J. et al., U.S. Pat. 4,024,043.

9   Chang, Y. (Ph.D. Thesis, Columbia University, New York) 1979.

10 Simons, R., Nature **280**, 30 (1979) 824.

11 Simons, R., Desalination **28** (1979) 41.

12 Grossman, G. et al., Desalination **10** (1972) 157.

13 Simons, R., Electrochim. Acta **30**, 1 (1985) 275.

Further Readings

1   Helfferich, F., *Ion Exchange* (McGraw-Hill, New York) 1962.

2   Katz, W., Electrodialysis preparation of boiler feed and other demineralized waters. Proc. Am. Power Conf. **33** (1971) 830-40; Electrodialysis for low TDS waters. Ind. Water Eng. June/July (1971).

3   Lakshminarayanaiah, N., Chem. Rev. **65** (1965) 491.

4   Mason, E. et al., Design of electrodialysis equipment. Chem. Eng. Prog. Symp. Ser. **55** (1959) 173-89.

5   McRae, W. "Electrodialysis" in *Kirk-Othmer Encyclopedia of Chemical Technology*, 3rd ed., Vol. 8 (Wiley, New York) 1978-1984, pp. 725 et seq.; "Electrodialysis" Chap. 8 in *Desalination Technology*, A. Porteous, ed. (Applied Science, London) 1983.

6   Shaffer, L. et al., "Electrodialysis" Chap. IV in *Principles of Desalination*, K.S. Spiegler, ed. (Academic Press, New York) 1966.

7   Solt, G., "Electrodialysis" Chap. 6 in *Membrane Separation Processes*, P. Meares, ed. (Elsevier, Amsterdam) 1976; "Electrodialysis" Chap. 12 in *Industrial Electrochemical Processes*, A. Kuhn, ed. (Elsevier, Amsterdam) 1971.

8   Spiegler, K. "Electrodialysis" in *Perry's Chemical Engineers' Handbook*, R. Perry et al., eds., 6th ed. (McGraw-Hill, New York) 1984, pp. 17-37.

Discussion

Chairman: H. Strathmann
Fraunhofer Institut, Stuttgart, West-Germany

G.H. Gessinger • To my knowledge, electrodialysis is still an emerging technology and I was therefore very surprised to hear that in the People's Republic of China there are 2000 plants in commercial use. What was the justification for such a large and early introduction of this technology in China?

W. McRae • The reason for introducing some desalting equipment was of course that they had a need in many of the provinces of China for potable water. The water supplies were saline either because of intrusion of seawater into groundwater or because in the western part of the country, many water supplies were brackish. They began development of electrodialysis around 1960 and electrodialysis was by far the easiest kind of process for them to develop. The development of reverse osmosis in the western world was really due to the efforts of one man, the late K.C. Channabasappa of the Office of Water Research and Technology in the US who demanded, with the funds of the US Government behind him, that reverse osmosis be developed. He did not give any contracts to the People's Republic of China and so they developed electrodialysis.

J. Stankovic • Could you please, in connection with this question, tell us a little bit more about the competition between electrodialysis and reverse osmosis? Electrodialysis is certainly very competitive for brackish water, especially in countries where the use of chemicals is problematic, which is the case for African countries and in the Middle East. Could you tell us some field where either electrodialysis or reverse osmosis are clearly favoured?

W. McRae • I think there are several reasons for choosing one above the other. If you use reversing type electrodialysis then you have no chemical feeds and that's a very good thing in developing countries, i.e. when the costs of chemicals are very high. In the Arab countries for example, chemicals are very highly priced but electricity is not. That tends to make electrodialysis economically superior to reverse osmosis even in the brackish water range. On the other hand, there are basic differences between RO and ED. Electrodialysis takes out the minor component (salt) from the major component (water). In electrodialysis, if you've got water which is already essentially potable (99.9% water, 0.1% salt) and you want to pretreat it before an ion-exchange bed, then that's a very inexpensive thing to do. The amount of energy is very small. In the brackish water range, the two processes are competitive and it depends on local circumstances, i.e. what's the cost of electricity and what are the costs of chemicals. In the range from say 2,000 ppm to 10,000 ppm, ED and RD compete against each other and you can't tell which one is going to win, except by doing the detailed design study. Another area is making very concentrated solutions, for example, making salt from seawater where the osmotic pressure differences are enormous (200 atmospheres or so). Then electrodialysis is again the economic way because the energy involved to transport the ions across the membranes is independent of concentration. As I said, the energy involved is in the order of magnitude of a quarter to half a kWh per kg of salt removed.

M. Kavanaugh • Mr. McRae, you mentioned the fact that ED is not used for nitrate removal. There is a widely recognized problem in parts of the United States particularly in

Californa due to agricultural drainage problems. Could you comment on why this has not occurred?

W. McRae • I think electrodialysis for removing and concentrating nitrate is a very feasible process. There just is not really a demand. In some European countries, without naming any names, it has been easier to change the regulations for nitrate content than to remove the nitrate.

R. Rautenbach • May I comment on this because we did some research on a pilot plant scale, comparing reverse osmosis and electrodialysis. Especially for the removal of nitrate, electrodialysis with polarity reversal is very attractive because the recovery rate which can be achieved is very high. Operating with polarity reversal, the concentrate can be supersaturated without scaling. Combining this with a crystallizer, we achieved 99.9% water recovery. This cannot be achieved by reverse osmosis because of scaling. That is an example where electrodialysis proves very interesting. The main reason why nitrate removal processes are not really applied is that the nitrate can be reduced to the acceptable limit by mixing sources of higher contents with water of a lower content. This is how waterworks can avoid these processes and still meet the official standards.

W. McRae • Sometimes waterworks ignore the official standards. It is easier. I might comment on the question of super-saturation in electrodialysis. It is true that using the reversing type of electrodialysis one can operate with Langelier indexes in the brine stream of around 2. That means that the actual pH in the brine stream is 2 pH units above the point at which calcium carbonate would precipitate. Nevertheless, you do not get scales of calcium carbonate. You can also operate at 200% saturation in calcium sulphate without using any chemical addition whatsoever. And if you are willing to add a few ppm of, say, sodium hexa-meta phosphate you can get up to 400% calcium sulphate saturation in the brine stream without any scale formation.

G. Kreysa • My question is related to the demineralization of water by electrodialysis. What, in your opinion, is a reasonable lower limit of salt concentration in the product stream, taking into account the fact that at very low salt values, the resistance and hence the cell voltage become unreasonably high?

W. McRae • I would say that your question is not quite accurate. It is true that the resistance of the solution becomes very high at low concentrations, but the reason the resistance is high is because there is not much salt there. And that's why the rule of thumb, half a kWh per kg of salt removed, still holds. So it isn't that the power consumption increases because the resistance goes up. In fact, the power consumption per m^3 of water produced goes down. The limitation is really a trade-off on investment costs between electrodialysis and say ion-exchange. In my opinion, in producing ultrapure water, one should demineralize to the order of magnitude of 30 to 50 ppm by electrodialysis and then finish by using a mixed-bed deionizer.

G. Schock • You mentioned several applications. One was the treatment of cooling water blow-down. My question is: what recoveries can be achieved and what is done with the concentrate in this application?

W. McRae • I think the best way to put it is that it's quite easy to get brine streams of 15 to 20% dissolved solids. I cannot answer the question of what the recovery is because you then have to tell me what salinity you are limited to in your cooling tower. For disposal some kind of an evaporation process is needed, e.g. an evaporation pond or, as we were saying some minutes ago, using a vapor compression evaporator to crystallize out the salt. If you use the kind of scheme that Prof. Rautenbach was talking about, you can get an overall recovery of 99.9%.

J. Hoigné • First of all a short comment regarding the denitrification of drinking water. I think, here we are in a very technocratically minded group, which is why we always think about denitrification. But I think we also should think about preventing the nitrate from getting into the groundwater. But anyhow, I still have a question about denitrification. When reverse osmosis is used, a secondary problem is that water is produced which is out of the carbonate equilibrium. This has to be re-established in a subsequent step. Now my question: are there electrodialysis membranes which are selective for nitrate and inhibit the transport of carbonate ions? This would be very advantageous for the treatment of drinking water.

W. McRae • Anion exchange membranes are normally quite selective for nitrate. The selectivity could probably be optimized, but it's pretty good right now. We don't take out much bi-carbonate when we take out nitrate.

Reactor Design for Electrochemical Water Treatment

G. Kreysa

Dechema Institut, Frankfurt, West Germany

1. Introduction

Waste water containing toxic metal ions is produced in many industrial branches, such as electroplating, cellulose acetate production, photographic development, printed circuit and battery production. A classical process for purification of such metal containing waste water is neutralization combined with metal hydroxide precipitation. Stricter legal regulation of effluent pollution calls for new, reliable and cost-effective processes for the purification of waste water containing heavy metals. The maximum allowable metal concentrations of various metals in effluents and the minimum metal concentrations obtainable by hydroxide precipitation at pH 8 are listed in Table 1.

These data show that only in a few cases is metal hydroxide precipitation able to attain the small metal ion concentrations which are required by waste water regulation laws in the countries shown. Furthermore, the disposal of metal hydroxide sludges is becoming increasingly expensive. Additional costly measures are required in order to avoid contamination of ground water with metal ions leached out by rain-water.

As an alternative to hydroxide precipitation, the ion-exchange technique is gaining in importance[2]. This method is, however, too costly for many types of effluents and regenerable ion-

Table 1:   Effluent limits compared with metal concentrations obtained by hydroxide precipitation[1].

Metal	Effluent limits in ppm			Metal concentration after precipitation at pH 8 in ppm
	D	USA	CH	
Pb	2	0.5	0.5	21
Cd	0.5	0.3	0.1	1500
Cu	2	0.5	0.5	1
Ni	3	0.5	2	340
Hg	0.05	-	0.01	-
Ag	2	-	0.1	-
Zn	5	0.5	2	2.6
Sn	5	-	2	-

exchange resins are not available for all metals, hence the incentive to develop electrochemical processes for pollution control. During the last decade, many electrochemical processes based on cathodic electro-deposition of metals have been developed and commercialized. These are being used increasingly in industry[3-11].

As a consequence of Nernst's law, the final equilibrium concentration of metal ions

$$C_{Me^+} = \exp\left[\frac{v_e F}{RT}(E - E_0)\right] \tag{1}$$

may be as small as one likes if the electrode potential E is maintained sufficiently low. Lowering the electrode potential, however, is limited practically by cathodic hydrogen evolution which decreases the current efficiency for metal deposition at very low electrode potentials. In Fig.1, a typical cathodic polarization curve of a metal containing waste water is shown illustrating this situation. In order to have available a large potential range for cathodic deposition of metals, electrode materials should be employed which have a high overvoltage for hydrogen evolution. Suitable electrode materials are e.g. graphite, lead or zinc.

For the usually small concentrations of metal ions in waste waters, the metal electro-deposition reactions are nearly always diffusion controlled. Due to low metal concentrations in effluents (some 100 ppm) previously mentioned as well as Faraday's law, the electricity costs for electrochemical waste water purification are quite low, i.e. usually in the order of 0.01 – 0.05 DM/m³. More important are the investment costs. Specific investment costs are determined by space-time yield given by the following formula[9,12]:

$$\rho = a_e \frac{M \Phi^e i}{v_e F} \tag{2}$$

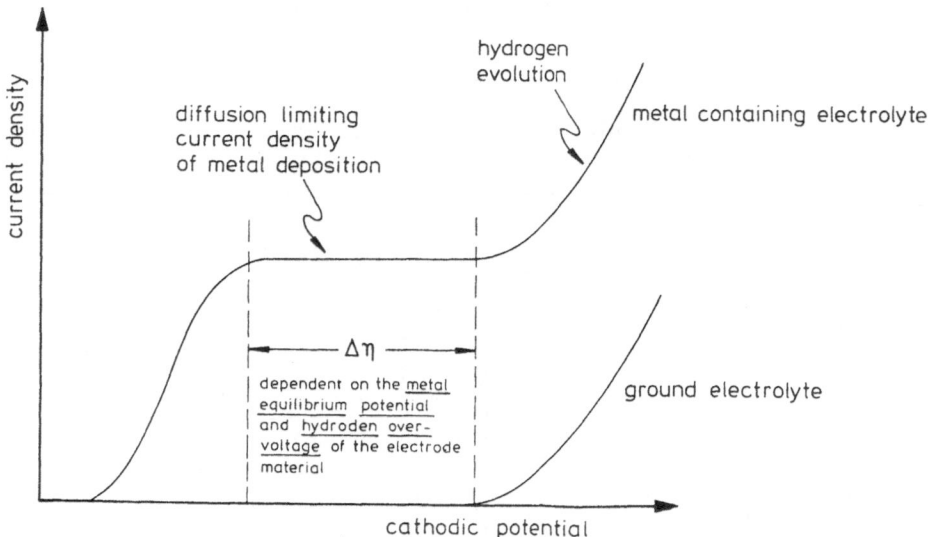

Fig. 1: Schematic cathodic polarization curve of a metal-containing waste water.

Introducing the equation for the diffusion controlled limiting current density

$$i = k \nu_e F c \tag{3}$$

yields the key formula for the design of electrochemical waste water purification reactors showing that at low concentrations a high specific electrode area a_e and a large mass transfer coefficient k is required:

$$\rho = \Phi^e M a_e k c \tag{4}$$

The space-time yield is the amount of metal which can be deposited within 1 cm³ of cell volume during 1s. However, space-time yield is dependent on inlet concentration and degree of conversion and thus does not allow a comparison of reactor performance independent of waste water properties. The introduction of a normalized space velocity has therefore been suggested[12,13] to characterize the performance of electrochemical waste water cells:

$$\rho_s^n = \frac{I \Phi^e}{(c_i - c_e) v_R v_e F} \log (c_i / c_e) \tag{5}$$

This normalized space velocity gives the volume of waste water in cm³ for which the concentration of the impurities can be reduced by a factor of 10 during 1s in a reactor volume of 1 cm³. A more appropriate dimension l/lh is given by multiplication with $3.6 \cdot 10^3$.

2. Industrial Electrochemical Reactors and Applications

As already mentioned, an efficient cell design for electrochemical waste water treatment requires a high specific electrode area and a large mass transfer coefficient. With respect to these criteria, electrochemical processes for the purification of effluents may be classified into three groups:

- The mass transfer rate and the current density can be enlarged by setting the electrodes in motion or by applying turbulence promoters. Examples are the pump cell[14,15], the Chemelec cell[16,17], the ECO[8,18-20], and the beat rod cell[21,22], and cells with vibrating electrodes or electrolytes[23].

- Attempts to accommodate large electrode areas in a small cell volume have resulted in developments such as the multicathode cell[24], the Swiss-roll cell[25,28], and the ESE cell[29].

- High mass transfer coefficients and large specific electrode areas are commonly obtained by using three dimensional electrodes. Examples are the porous flow-through cell[30-33], the RETEC cell[34], the packed-bed cell[8,35-37], the fluidized-bed cell[38-42], and the rolling tube cell[21,22].

The following discussion is confined mainly to those cell types which are used in industrial applications.

A cell suitable for smaller applications is the so-called pump cell[14,15] which is a rotating analogue to the capillary gap cell[43]. This cell is shown schematically in Fig. 2. It consists of two stator disc electrodes which act as endplates with electrical connections. In between, on a rotating shaft, a bipolar rotating disc electrode is mounted. The electrolyte flows radially from the central tube to the outer circumference of the rotating electrode. Due to the rotation of the

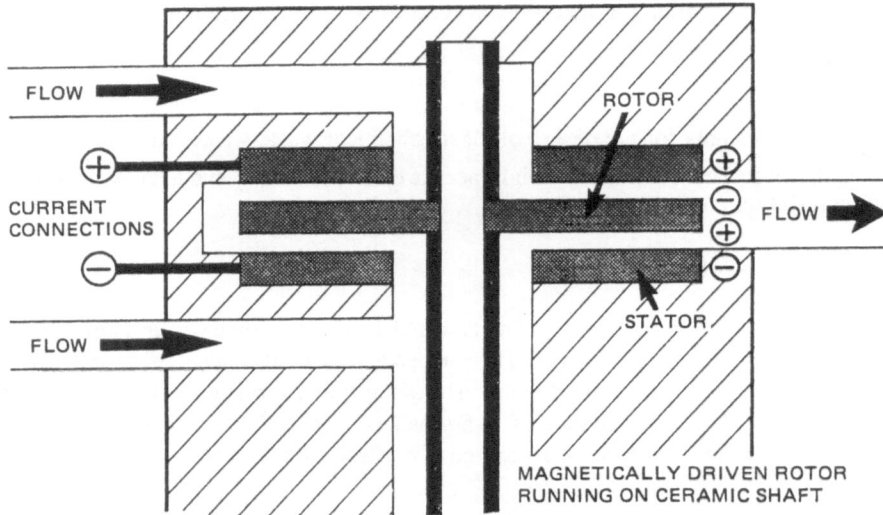

SCHEMATIC DIAGRAM OF BIPOLAR PUMP CELL (NOT TO SCALE)

Fig. 2: Schematic view of the pump cell.

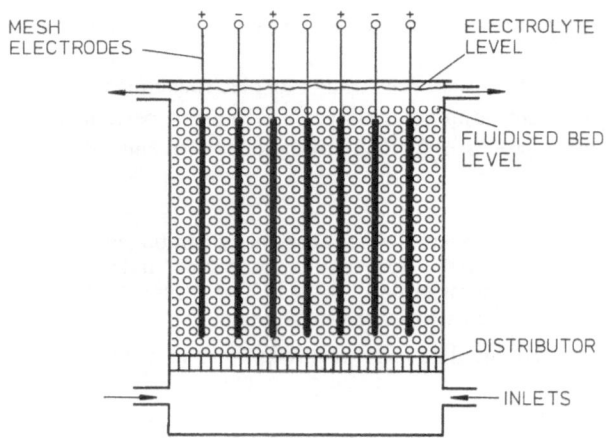

Fig. 3: Sketch of the Chemelec cell.

electrode, a high mass transfer coefficient can be obtained. A special advantage of this cell is that the mass transfer coefficient and the residence time is only determined by the radial flow rate of the electrolyte whereas the mass transfer coefficient is given mainly by the rotation speed. Apart from some laboratory experiments, this cell has not yet been employed on an industrial scale.

Another way of improving the mass transfer to a plane electrode is to use a fluidized bed of glass spheres as a turbulence promoter. The Chemelec cell[16,17], which is shown schematically

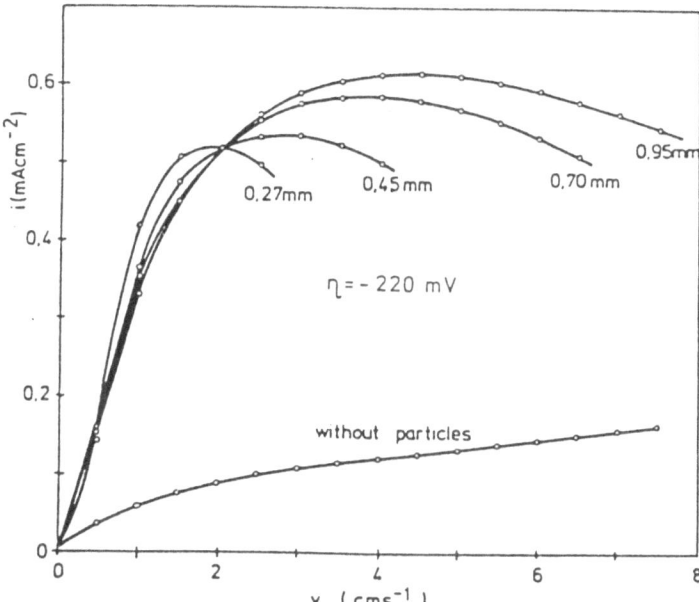

Fig. 4: Limiting current density of hexacyanoferrate reduction in a Chemelec cell as a function of flow velocity for different particle diameters and without particles[44].

in Fig. 3, uses this technique. The electrolyte enters the cell at the bottom via an electrolyte distributor plate. The cell tankhouse contains several monopolar plate electrodes and glass spheres of about 1 mm diameter which are fluidized by the electrolyte flow. That mass transfer is increased using this system can be seen from the experimental data given in Fig. 4. The diffusion limited current density for the reduction of hexacyanoferrate at a plane electrode in a channel has been measured as a function of flow rate in the absence of a fluidized bed and in the presence of a fluidized bed consisting of particles of various diameter[44]. Compared with the laminar flow parallel to a plate electrode, the mass transfer coefficient is enhanced by up to a factor of 6 in the presence of a fluidized bed acting as turbulence promoter.

Since the electrolyte flow velocity has to exceed the minimum fluidization velocity, the residence time and also the degree of conversion per pass are limited. This cell is therefore suitable for a pre-treatment or for a recycling loop operation mode. In electroplating industries, the cell can be used to remove metal ions from rinsing waters. In the steady-state, metal is deposited in the cell at the same rate as it is introduced into the bath. By means of this process it is possible to maintain the copper concentration at a 200 ppm level in the primary rinsing bath of an electroplating line.

Another electrode configuration that achieves high mass transfer coefficients is the rotating cylinder electrode[18]. The so-called ECO cell, which was designed on this principle in the U.K.[19,20], has also been industrially produced. The cascade configuration shown in Fig. 5 is necessary for applications involving not only metal recovery but also effluent purification. The reason for this is that only a limited degree of conversion can be obtained in one pass with a single cell. In the cascade cell, a single rotating cylinder of about 50 cm diameter serves as a cathode for all chambers. Partition walls are provided to subdivide the catholyte chambers into

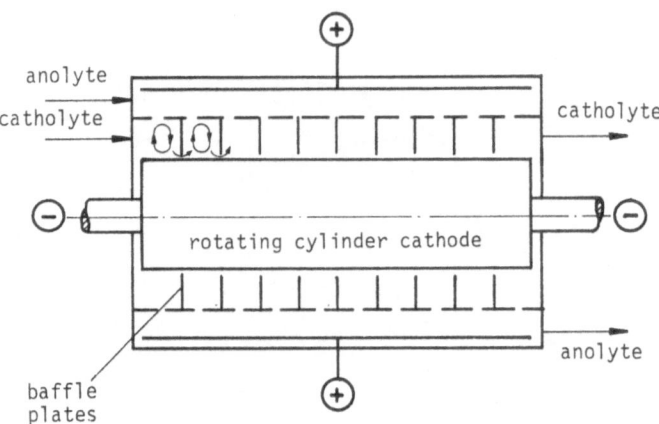

Fig. 5:   Sketch of the ECO-cell cascade configuration.

Table 2:   Operation data for copper removal by an ECO-cell cascade (1.7 m²,
             12 compartment cascade cell).

Inlet concentration	100 ppm Cu^{2+}
Outlet concentration	2 ppm Cu^{2+}
Throughput	8 m³/h
Current	1000 A
Mean c.d.	59 mA/cm²
Current efficiency	0.65
Power consumption	
• electrolysis alone	1.5 kWh/m³
• including rotation	3 kWh/m³

segments through which the waste water successively flows. This largely eliminates back-mixing effects so that high degrees of conversion can be achieved resulting in the required low final concentrations. Some typical operating data of the ECO-cell cascade for copper removal are summarized in Table 2.

A special cell design, mainly used in small electroplating shops, is the beat rod cell[20,21] shown in Fig. 6. Cathode rods are supported in an electrolysis tank and are slowly rotated within an annular type of chamber. The deposited metal can be made to settle as a powder by causing the rods to strike one another and can be discharged at the bottom of the cell. This reactor can only be operated batch-wise, thus restricting its use to small throughputs. A typical concentration depletion curve for silver deposition in the beat rod cell is shown in Fig.7.

Another interesting way of enhancing the mass transfer rate in a parallel plate tank cell is to apply mechanical vibrators to the cell housing[23]. This allows vibrations to be transferred to the electrolyte resulting in the increase of the mass transfer coefficient. Such cells are used in industry, especially if higher metal contents have to be treated and if the required conversion degree is not too high. A typical application is the pretreatment of waste water, which is finally

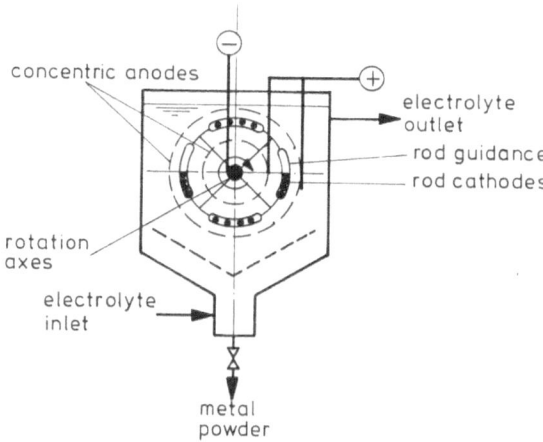

Fig. 6: Schematic view of the beat rod cell.

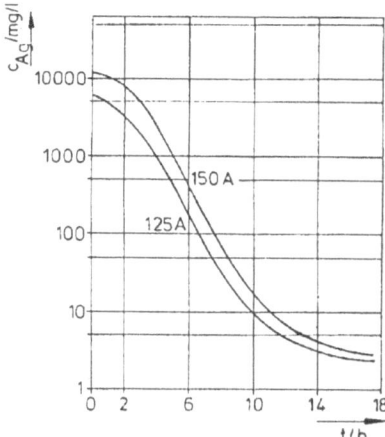

Fig. 7: Concentration-time curves for silver removal by the beat rod cell[21,22]

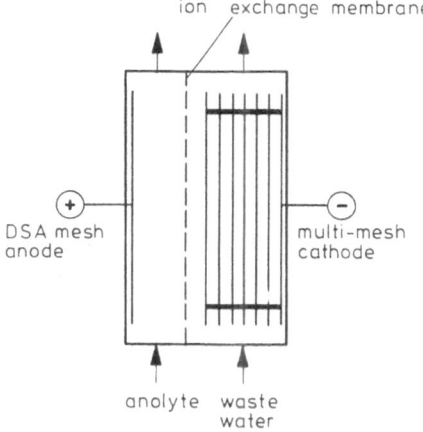

Fig. 8: Sketch of the multi-cathode cell containing a cathode stack of expanded metal sheets.

purified by packed bed electrolysis. This is discussed in more detail below.

Another cell design intended to accommodate a larger electrode area in a small cell volume is the multi-cathode cell shown schematically in Fig. 8 [27,45]. (This has been used for gold recovery[45].) Based on the usual configuration of planar electrodes, the single cathode is replaced by a stack of expanded metal sheets. The application of a stack of meshes for laboratory cells is also described in the literature[46,47]. Ohmic losses, however, occur inside such stack structures, restricting the penetration depth of the current[8,9]. Depending on the operating parameter of the

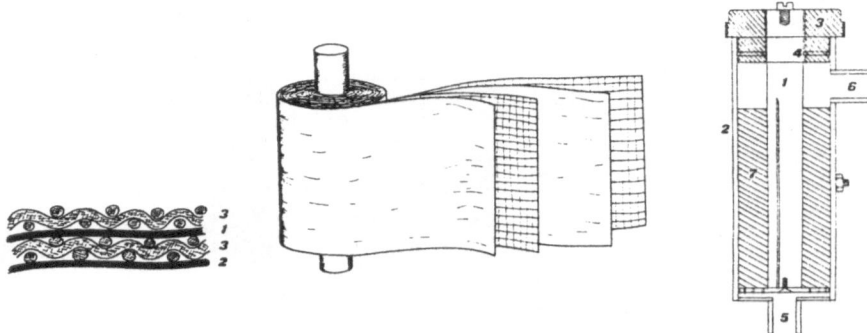

Fig. 9:   Sketch of the Swiss-roll cell.

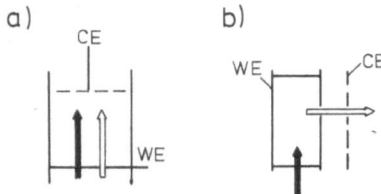

Fig. 10:   Diagram of flow-through (a) and flow-by (b) arrangement of cells with
               3D electrodes, CE:counter electrode, WE: working electrode.

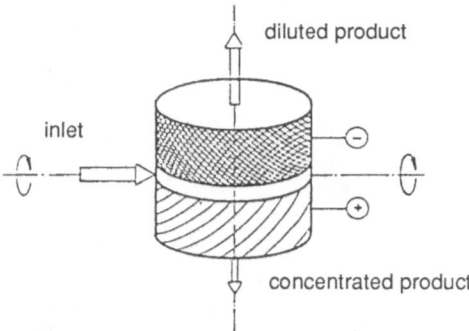

Fig. 11:   Example of porous flow-through electrolysis for concentration of dilute
               solutions[30].

electroysis, only the first 5 to 10 electrodes facing the diaphragm are electrochemically active. This restricts the thickness of the electrode stack and limits the scaling up.

An efficient way of fitting larger planar electrode surface areas in a small volume and thereby avoiding the problems mentioned in connection with the multi-cathode cells is used in the Swiss-roll cell shown in Fig. 9 [25-28]. In this design, metal foil sheets separated by an electrolyte permeable medium (e.g. plastic mesh) are wrapped helically around a core. The voltage drop

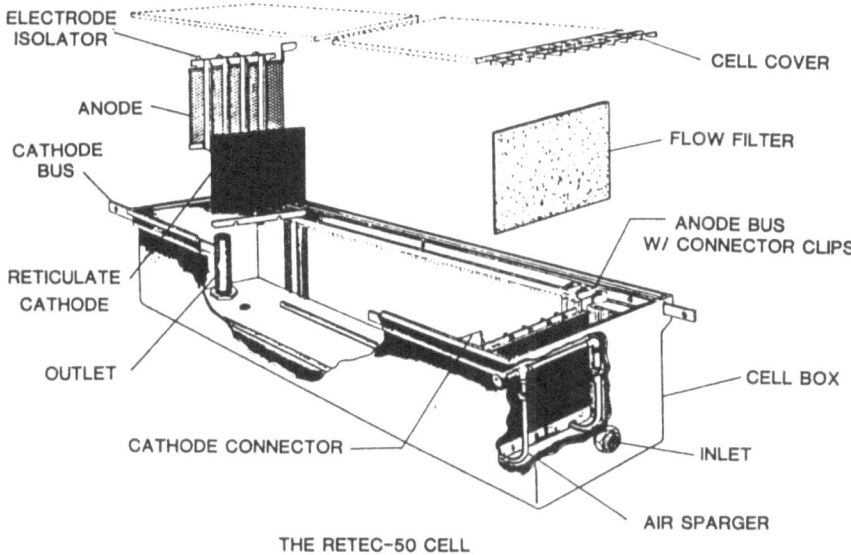

ELECTRODE
ISOLATOR

CELL COVER

ANODE

FLOW FILTER

CATHODE
BUS

ANODE BUS
W/ CONNECTOR CLIPS

RETICULATE
CATHODE

OUTLET

CELL BOX

CATHODE CONNECTOR

INLET

AIR SPARGER

THE RETEC-50 CELL

Fig. 12:   Exploded view of the RETEC cell[34].

which occurs in the thin foil is largely compensated for by connecting up the electrode electrically at opposite ends instead of at the same ends[48]. The effluent flows axially through the electrode pack housed in a tubular cell and the metals are removed by being deposited on the cathode foil. When the cell is filled with metal after a deposition period of about 1 to 2 hours, the latter is chemically dissolved in acid and removed in a regeneration step. Since the cell has no diaphragm, it is only suitable for reactions which can be carried out in an undivided cell. However, in general, this does not present a problem when handling metal-containing effluents of simple composition. Large space time yields are obtained with the Swiss-roll cell due to its large specific electrode area. The mass transfer conditions are also favourable since the separator functions simultaneously as a turbulence promoter. The capacity of the cells can be increased by replacing the cathode foil with layers of mesh electrodes[27], thereby achieving a truly three-dimensional electrode.

Another design, which is very similar to the Swiss-roll cell, uses mesh electrodes instead of foil electrodes and is known as the ESE cell (Extended Surface Electrolysis) in the literature[29]. On an industrial scale, this cell has been successfully used for the purification of effluents containing copper. Using mesh electrodes and a perforated tube as a winding core, a radial flow version[49] of the Swiss-roll cell is obtained.

A simple three-dimensional electrode is obtained when using packed beds of conductive particulate material through which the electrolyte flows. Numerous versions of these porous flow cells have been described in the literature[8,9,30-33,50-54]. Two principal arrangements of cells with three-dimensional electrodes are possible with respect to electrolyte and current flow directions. These are referred to in the literature as flow-through and flow-by arrangements. Both are illustrated in Fig. 10. An interesting example is shown in Fig. 11. The cell contains cylindrical packed beds for both anode and cathode[30]. The effluent is fed into the cell between the electrodes and enters the two electrodes through distributor plates (porous plastics). External valves are used to adjust the flow rates in such a way that about 99% of the effluent

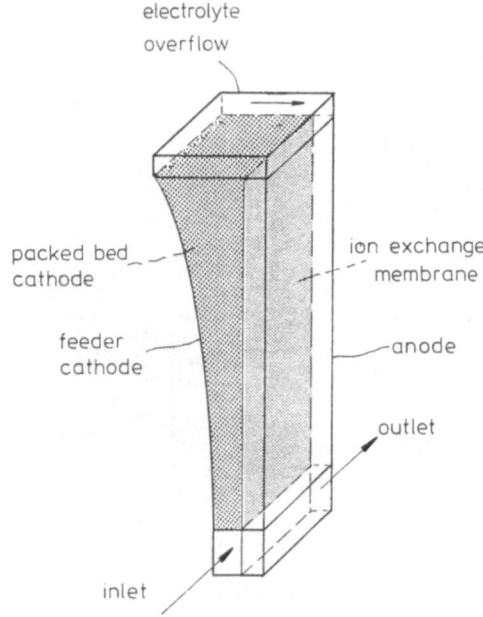

Fig. 13:   Schematic representation of the Enviro cell.

Table 3:   Some industrial applications of the Enviro cell[1].

Application field	Metal	Throughput m³/h	Inlet conc. ppm	Outlet conc. ppm	Energy consumption kWh/m³	Anode area m²
Production of measuring instruments	Hg	0.3	300	0.05	1.2	1
Film processing	Ag	0.2	15	1.0	0.15	1
Salt production	Pb	0.5	2	0.1	0.07	1
Electroplating	Cd	0.2	20	0.1	0.18	1
Battery production	Hg/Cd	0.08	500	0.01	1.7	3
Cellulose acetate production	Cu	20	20	1.9	0.08	40
Pickling (recycling of solution)	Cu	3	150	50	0.19	5
Dyestuff production	Cu	6	400	2.0	4.0	90
Dyestuff production	Hg	2	4	0.05	2.5	15

passes through the cathode and only 1% through the anode. When the cathode is filled with deposited metal, the whole unit is turned upside down so that the electrode loaded with metal becomes the anode. The metal is dissolved anodically and the anolyte stream then consists of a concentrated metal salt solution. This cell can therefore be considered as a continuously operating device which, in a simple manner, produces basically the same result as an ion exchanger.

A further quite efficient electrochemical system for waste water purification is the RETEC cell[34]. In principle this cell employs three-dimensional cathodes, but it is similar in design to a simple tankhouse cell. The cathodes consist of flow-through metal sponge electrodes. These cathodes have an active surface area approximately 15 times their geometric area. An exploded view of a complete RETEC cell is shown in Fig. 12. Typical applications of this system are similar to those of the Chemelec cell. It is used in closed water recycling systems in the electroplating industry in order to maintain metal concentrations in rinsing water at a reasonably low level.

In three-dimensional electrodes, the penetration depth of the current in a direction parallel to the current flow is limited. Ohmic losses in the electrolyte are the main cause of the decrease in local current density. This approaches nearly zero with increasing distance from the counter-electrode[8,9,50-54]. For a diffusion controlled reaction, the penetration depth of the limiting current density can be calculated using the following formula:

$$h_p = \left[\frac{2 v \kappa_s \Delta\eta}{a_e k v_e F c} \right]^{0.5}$$ (6)

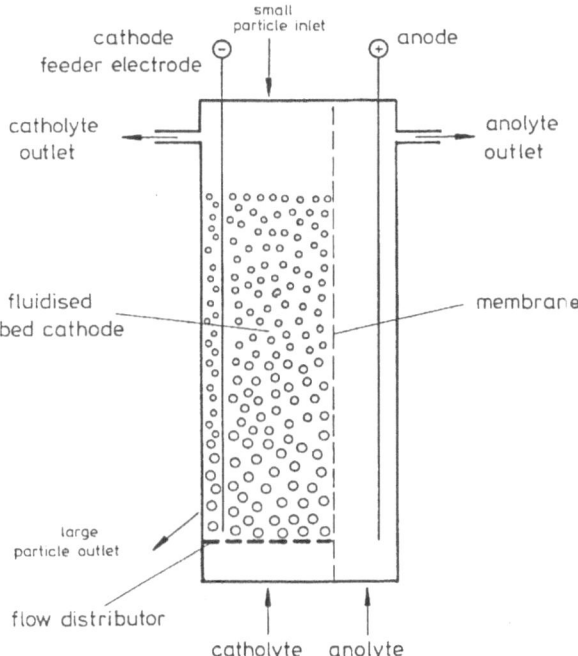

Fig. 14:   Sketch of a fluidized bed electrolysis cell.

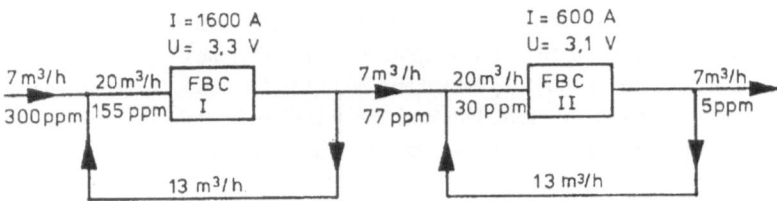

Fig. 15:   Flow sheets for copper removal by fluidized bed electrolysis.

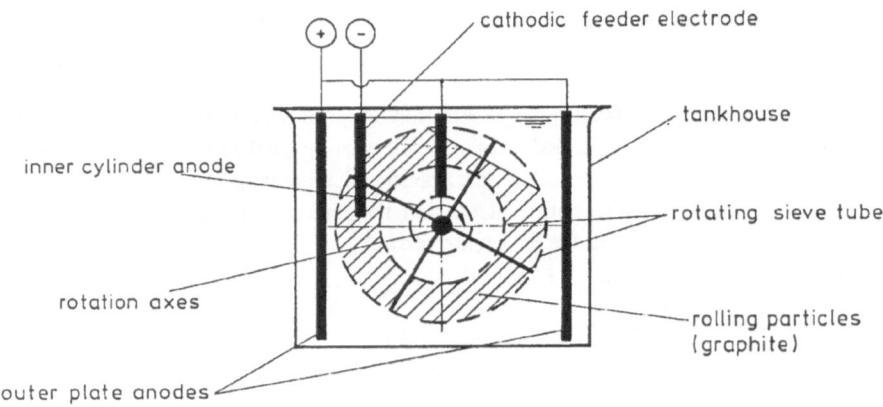

Fig. 16:   Diagram of the rolling tube cell[21,22].

Table 4:   Operation data for different waste water electrolysis cells.

Process	Scale	Re-moved metal	C_0 [ppm]	C_1 [ppm]	V_R [l]	V_D [lh^{-1}]	I [A]	i [Acm^{-2}]	U_z [V]	ϕ^e
ECO cell	Ind.	Cu	100	2	275	8000	1000	0.059	12	0.65
Beat rod cell	Ind.	Ag	6000	4	200	12.5	125	-	5	0.15
Swiss-roll cell	Lab.	Cu	380	25	0.3	0.7	0.55	0.46•10^{-3}	1.56	0.38
Porous flow-through cell	Lab.	Cu	800	0.2	2.6	0.64	0.45	3.5•10^{-3}	1.46	0.95
Packed bed cell	Ind.	Cu	50	0.1	4.8	50	3.2	0.76•10^{-3}	1.9	0.66
Fluidized bed cell	Ind.	Cu	77	5	192	7000	600	-	3.1	0.71
Rolling tube cell	Ind.	Au	81	0.4	40	6.9	15	-	6.05	0.005

According to this equation, the penetration depth of the limiting current density increases with decreasing concentration. The application of this principle led to the design of the Enviro-cell which is shown schematically in Fig.13 [35-37]. According to equation (6), the depth of the bed increases as the metal ion concentration decreases from the lower inlet to the upper outlet. This cell has a high space time yield and, with residence times of only a few minutes, the metal concentration can be reduced by up to 1/1000 of the initial concentration.

The Enviro-cell has found a wide range of industrial applications. Some of them are given in Table 3.

The principle of fluidized-bed electrolysis designed by Goodridge and Fleischmann is shown in Fig. 14 [39]. The electrolyte flows from bottom to top through a loose bed of particles thus fluidizing it. The fluidized particles are charged cathodically via a feeder electrode. Metal ions are hence deposited cathodically on the particles. The particles grow and the larger ones collect in the lower part of the bed and can then be discharged. They are replaced by fresh small particles fed in at the top of the cell. Fluidized bed electrolysis has been examined and tested for different applications in many laboratories[38,55-62] and has also been employed on an industrial scale[40-42]. In a fluidized bed cell, a minimum flow rate corresponding to the minimum fluidization velocity must be exceeded. Since the overall height of such a cell is usually restricted to about 2 m for hydraulic reasons, only a limited residence time is possible. This means that only a limited concentration decrease per pass can be achieved. To overcome these problems, continuous operation with recirculation is used and several cells are connected in series. Fig. 15 shows a flow sheet of a typical industrial plant for the purification of effluents containing copper. The volumetric flow rates and copper concentrations of the individual streams as well as the electric data of the cells are indicated.

Another rotating design with a special fluidized bed electrode is described in the literature as a rolling tube cell[21,22] (see Fig. 16). This resembles in principle the well-known plating process for piece-goods using slowly rotating barrels. The rotating drum is only partially filled with graphite particles in order to achieve thorough agitation. The cathode takes the form of a perforated annular-type drum and is surrounded on the inside and outside by anodes in order to render the current distribution as homogeneous as possible. This cell has proved to be especially useful for the recovery of silver and gold.

To enable better comparison of all the different cell designs and their applications discussed above, a summary of their operating features is given in Table 4.

Table 5:   Comparison of various figures of merit for waste water electrolyis cells.

Process	E_s^e [kwh/m^3]	E_s^t [kwh/m^3]	ρ_s^n [l/lh]
ECO cell	1.5	3	20
Beat rod cell	50	60	0.2
Swiss-roll cell	1.23	-	20
Porous flow-through cell	1.03	-	0.9
Packed bed cell	0.12	-	28
Fluidized bed cell	0.27	-	30
Rolling tube cell	14.2	-	0.4

Since this data is obtained from cells of different sizes dealing with different kinds of effluents, they are intended to serve as a rough guide rather than as a reliable basis for comparison of the individual methods.

Another way of comparing the processes dealt with is given in Table 5. Here the data are closely correlated with economic efficiency.

The specific electric energy consumption per m^3 of effluent to be purified, E_s^e, is relevant for the calculation of energy costs. The total energy consumption E_s^t including that of a rotation mechanism is specified in addition for all cells containing rotating parts. An important criterion which strongly influences the total cost is the normalized space velocity ρ_s^n also given in Table 5 for typical applications. When comparing specific energy consumptions and normalized space velocities for processes already in industrial use, packed bed and fluidized bed electrolysis seem to perform best.

3. Plant Engineering Aspects

For the operation of industrial electrochemical waste water purification cells, several operations modes are available. The two principal alternatives for every continuous process are shown in Fig. 17. A reaction engineering analysis of both of these plant schemes must distinguish between the ideal cases of a stirred vessel and a tube reactor. The Enviro cell and the RETEC cell represent close to ideal tube reactors, whereas the other systems may be described as approximating stirred vessel reactors.

The basic differential equation for a continuous stirred vessel reactor under limiting current conditions is given as:

$$\frac{dc}{dt} = \frac{1}{\tau_R}(c_i - c) - a_e k c. \tag{7}$$

The complete solution including the time variable is given by:

$$c(t) = \frac{c_i}{\tau_R a_e k + 1}\left\{1 + \tau_R a_e k \exp\left[-\left(a_e k + \frac{1}{\tau_R}\right)t\right]\right\}. \tag{8}$$

The residence time τ_R is given as v_R/q. Some concentration-time curves calculated using this equation are shown in Fig. 18. For infinite time, equation (8) transforms into the steady-state solution:

$$c_s = \frac{c_i}{\tau_R a_e k + 1}. \tag{9}$$

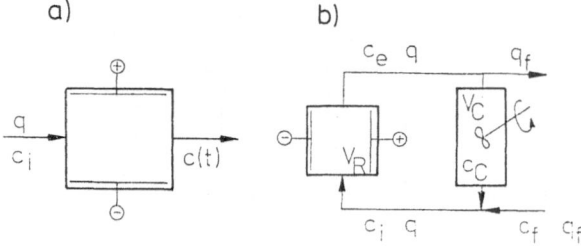

Fig. 17: Flow sheets of (a) continuous flow-through and (b) recyle operation modes of waste water cells.

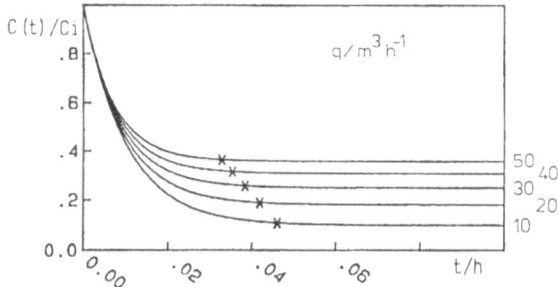

Fig. 18:   Concentration-time curves for the continuous stirred vessel reactor ($a_e = 5$ cm $^{-1}$, k = 5 x 10^{-3} cm/s, V_R = 1 m³).

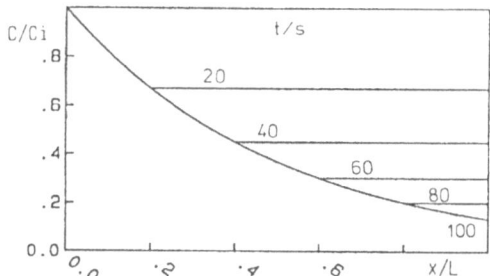

Fig. 19:   Time dependence of the concentration profile within an ideal tube reactor ($a_e = 20$ cm^{-1}, k = 1 x 10^{-3} cm/s, L = 50cm, u = 0.5 cm/s).

According to this equation, the lower the steady-state concentration, the lower the volumetric flow rate is, which means the higher the residence time is (see Fig. 18). However, not only is the steady-state itself influenced by the residence time, but so also is the time which is necessary to reach the steady-state. The crosses in Fig. 18 mark the time which is necessary to reach 99% of the steady-state. This shows clearly that the process dynamics slow down as the degree of conversion increases.

For an ideal tube reactor, the concentration within the reactor is a function of time and place (a site-coordinate parallel to the electrolyte flow). The problem is described by the following partial differential equation:

$$\frac{\partial c(t, x)}{\partial t} = - u \frac{\partial c(t, x)}{\partial x} - a_e k \, c(t, x) . \tag{10}$$

The solution of this equation is given by:

$$\frac{c(t, x)}{c_i} = [1 - U(x - ut)] \exp\left(-\frac{a_e k}{u} x\right) + U(x - ut) \exp(-a_e k t) \tag{11}$$

with the step function
$$U(x \leq 0) = 0$$
$$U(x > 0) = 1.$$

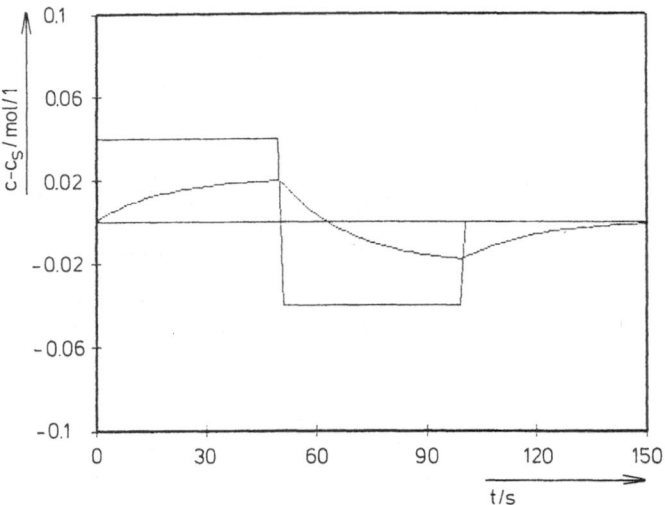

Fig. 20:   Damping of concentration distrubances by the stirred vessel reactor.

Fig. 19 shows the time dependence of the concentration profile within an ideal tube reactor calculated using equation (11). A steady-state profile is achieved within the residence time. At times shorter than the residence time, the front part of the reactor is already stationary while the concentration at the end continues to decrease.

It is also interesting to analyze an electrochemical reactor by means of dynamic system analysis. This method allows one to investigate how the steady-state outlet concentration changes if a concentration disturbance occurs at the reactor inlet. The principal effect is illustrated in Fig. 20. If, at the reactor inlet a rectangular concentration disturbance is applied, the concentration-time curve at the reactor outlet is damped, as is shown in Fig. 20 for a stirred tank reactor. This behaviour is quite important for technical applications because the metal ion concentration of an industrial waste water stream very often varies with time.

As an example of the behaviour of a continuous recycle loop operation, a stirred tank reactor is shown. The behaviour of such a system is determined by the following set of differential equations:

$$\frac{dc_e}{dt} = \frac{1}{\tau_R}(c_i - c_e) - a_e k c_e \qquad (12)$$

$$\frac{dc_i}{dt} = \frac{1}{\tau_C}\left(1 - \frac{q_f}{q}\right)c_e - c_i + \frac{q_f}{q}c_f \qquad (13)$$

with the boundary condition: $c_e(0) = c_i(0) = c_0$.   (14)

The steady-state solution is given by:

$$c_0 = \frac{c_f}{\tau_f \, a_e \, k + 1}, \quad \tau_f = \frac{v_R}{q_f}. \tag{15}$$

This result is quite similar to equation (9) for the continuous stirred tank reactor. However, the inlet concentration of the reactor and residence time of the reactor in equation (9) are here replaced by the corresponding values for the inlet stream at the edge of the loop (see Fig. 17). This means that the final concentration for a continuous recycle operation is always lower than for the continuous flow-through operation of a single reactor. A continuous recycle operation mode is very appropriate for all technical applications in waste water purification where the required degree of conversion cannot be obtained by a single pass of the electrolyte through the cell.

4. Conclusion

Due to strict legal regulations in almost all countries, today reliable and cost-effective industrial processes for removal of heavy metal ions from waste waters are required. In general, electrochemical cathodic deposition of the metals is a method suitable for this purpose. However, from an economic point of view the resulting low space-time yield at the usually low concentration levels is a problem. Therefore, several new electrochemical waste water purification processes have been developed during the last decade improving the space-time yield either by applying turbulent flow conditions or electrode design concepts offering a large specific electrode area. Some processes like the packed bed electrolysis (Enviro cell), the Chemelec cell, the ECO cell and others have already found a wide range of applications in different branches of industry. Reliable mathematical models for designing such cells and plants for a given application are available today.

Nomenclature

a_e	specific electrode area	(cm^2/cm^3)
c	concentration	$(mole/cm^3)$
c_e	exit concentration	$(mole/cm^3)$
c_f	feed concentration	$(mole/cm^3)$
c_i	inlet concentration	$(mole/cm^3)$
E	electrode potential	(V)
E_o	equilibrium potential	(V)
F	Faraday number	$(As/mole)$
i	current density	(A/cm^2)
I	current	(A)
k	mass transfer coefficient	(cm/s)
M	molar mass	$(g/mole)$
q	volumetric flow rate	(cm^3/s)
q_f	volumetric feed flow rate	(cm^3/s)
R	gas constant	$(Ws/K\,mole)$

　　　　　　　　　　　　　　G. Kreysa

t	time	(s)
T	temperature	(K)
u	linear flow velocity	(cm/s)
v	voidage	(1)
$\Delta\eta$	overvoltage range of limiting c.d.	(V)
κ_s	solution conductivity	(s/cm)
ν_e	electron number	(1)
ρ	space-time yield	(g/cm^3 s)
τ_R	reactor residence time	(s)
τ_C	container residence time	(s)
Φ^e	current efficiency	(1)

References

1　Müller,K.-J. and Kreysa, G., Dechema-Monographien **98** (1985) 367.

2　Kammel, R. and Lieber, H.-W., Galvanotechnik **68** (1977) 413.

3　Kuhn, A.T., Chem. and Ind. (1971) 946.

4　Flett, D.S and Pearson, D., Chem. Ind. (1975) 639.

5　Kammel, R. and Lieber, H.-W., Galvanotechnik **68** (1977) 883.

6　Kammel, R. and Lieber, H.-W., Galvanotechnik **69** (1978) 317, 624.

7　Kuhn, A.T., Chem. Ind. (1978) 447.

8　Kreysa, G., Chem.-Ing.-Tech. **50** (1978) 332.

9　Kreysa, G., Metalloberfläche **34** (1980) 494.

10　Fabjan, C., Oberfläche-Surface **21** (1980) 283.

11　Samhaber, W., Chem.-Ing.-Tech. **56** (1984) 246.

12　Kreysa, G., J. Appl. Electrochem. **15** (1985) 175.

13　Kreysa, G., Electrochim. Acta **26** (1981) 1693.

14　Jansson, R.E.W. and Marshall, R.J., Chem. Engineer **315** (1976) 769.

15　Jansson, R.E.W. and Tomov, N.R., Chem. Engineer **327** (1977) 867.

16　Lopez-Cacicedo, C.L., The Inst. Chem. Engs. Symp. Ser. No. 42 (1975) 29.

17　Lopez-Cacicedo, C.L., Brit. Pat. 1423369 (1973).

18　Gabe, D.R., J. Appl. Electrochem. **4** (1974) 91.

19　Holland, F.S., Chem. Ind. (1978) 453.

20　Ricci, L.J., Chem. Eng. (1975) 29.

21　Kammel, R. and Lieber, H.-W., Galvanotechnik **69** (1978) 687.

22　Götzelmann, W., Galvanotechnik **70** (1979) 596.

23　Bruhn, D., Dietz, W., Müller, K.-J. and Reynvaan, C., EPA 86109265.8 (1986).

24　Storck, A., Robertson, P.M. and Ibl, N., Electrochim. Acta **24** (1979) 373.

25　Robertson, P.M., Schwager, F. and Ibl, N., J. Electroanal. Chem. **65** (1975) 883.

26　Robertson, P.M. and Ibl, N., J. Appl. Electrochem. **7** (1977) 323.

27 Robertson, P.M., Scholder, B., Theis, G. and Ibl, N., Chem. Ind. (1978) 459.

28 Ibl, N. and Robertson, P.M. , Chem.–Ing.–Tech. **48** (1976) 165.

29 Keating, K.B. and Williams, J.M., Res. Rec. Conserv. **2** (1976) 39.

30 Bennion, D.N. and Newman, J., J. Appl. Electrochem. **2** (1972) 113.

31 Carlson, G.A., Estep, E.E. and Jacqueau, D., Chem.–Ing.–Tech. **45** (1973) 217.

32 Wenger, R.S. and Bennion, D.N., J. Appl. Electrochem. **6** (1976) 385.

33 Van Zee, J. and Newman, J., J. Electrochem. Soc. **124** (1977) 706.

34 Advertising information of ELTECH Electroresearch S.A., Genève.

35 Kreysa, G., Chem.–Ing.–Tech. **55** (1983) 23.

36 Kreysa, G. and Reynvaan, C., J. Appl. Electrochem. **12** (1982) 241.

37 Kreysa, G., DE 26 22 497 (1976).

38 Backhurst, J.R., Coulson, J.M., Goodridge, F., Plimley, R.E. and Fleischmann, M., J. Electrochem. Soc. **116** (1969) 1600.

39 Backhurst, J.R., Fleischmann, M., Goodridge, F. and Plimley, R.E., GB Pat. 1 194 181 (1970).

40 Scharf, H., DE 22 27 084 (1972).

41 Raats, C., Boon, H. and Eveleens, W., Erzmetall **30** (1977) 365.

42 v. Heiden, G., Raats, C. and Boon, H., Chem. Ind. (1978) 465.

43 Beck, F. and Guthke, H., Chem.–Ing.–Tech. **41** (1969) 943.

44 Kreysa, G., Pionteck, S. and Heitz, E., J. Appl. Electrochem. **5** (1975) 305.

45 W.C. Heraeus GmbH, advertising material, ACHEMA (1979).

46 Sioda, R.E., J. Electroanal. Chem. **34** (1972) 399.

47 Sioda, R.E., Electrochim. Acta **16** (1971) 1569.

48 Robertson, P.M., Electrochim. Acta **22** (1977) 411.

49 Gallone, P., De Anna, P.L. and Bonora, P.L., Materials Chemistry **3** (1978) 285.

50 Newman, J.S., Tobias, C.W., J. Electrochem. Soc. **109** (1962) 1183.

51 Chu, A.K.P. and Fleischmann, M., Hills, G.J., J. Appl. Electrochem. **4** (1974) 323.

52 Alkire, R. and Ng, P.K., J. Electrochem. Soc. **121** (1974) 95.

53 Bennion, D.N., J. Appl. Electrochem. **2** (1972) 113.

54 Trainham, J.A. and Newman, J., J. Electrochem. Soc. **124** (1977) 1528.

55 Flett, D.S., Chem. Ind. (1971) 300.

56 Flett, D.S., Chem. Ind. (1972) 983.

57 Steppke, H.-D. and Kammel, R., Erzmetall **26** (1973) 533.

58 Monhemius, A.J. and Costa, P.L.N., Hydrometallurgy **1** (1975) 183.

59 Kreysa, G., Erzmetall **28** (1975) 440.

60 Flett, D.S. and Pearson, D., Chem. and Ind. (1975) 639.

61 Germain, S. and Goodridge, F., Electrochim. Acta **21** (1976) 545.

62 Hutin, D. and Coeuret, F., J. Appl. Electrochem. **7** (1977) 463.

Discussion

Chairman: M. Mirbach
Brown Boveri, Baden, Switzerland

M. Mirbach   •   How important are organic impurities in waste water when you try to recover the metals?

G. Kreysa   •   In many of the applications I have shown you there are no problems with organic impurities because they are not there to begin with. In electro plating, however, you have almost always organic inhibitors in the baths, but we have not found that they cause any problems. It is also possible to direct the electrolyte flow first through the anode and then through the cathode chamber. In that way you may achieve partial oxidation of organic compounds resulting in a decrease in the chemical oxygen demand.

W.A. McRae   •   Have you done any studies of whether it makes sense to use reverse osmosis or electrodialysis in front of some of these apparatus to concentrate the metal waste?

G. Kreysa   •   We haven't done such studies at Dechema, but our licence partner and some customers have. The cells, especially those with three dimensional electrode designs, are optimized for relatively low inlet concentrations. If the inlet concentration is too high, the bed gets filled with metal rapidly and the recovery procedure has to be started at short intervals. In some cases, a pretreatment for reducing the inlet concentration is needed. For such a pretreatment we have found that cells like the Chemelec (see section 2), or a simple parallel plate cell , are much cheaper than electrodialysis or reverse osmosis. Combinations of two electrochemical systems are in use but I don't know of a combination with a membrane process.

S. Stucki   •   What processes do you consider to be the main competitors to the electrochemical removal of heavy metals?

G. Kreysa   •   I would say ion exchange.

S. Stucki   •   Is there a concentration, where ion exchange becomes more economic than the electrochemical reduction?

G. Kreysa   •   That's difficult to answer in a general way. Probably nearly half of all customers who have installed such plants have compared the process with ion exchange. If you have waste water containing not only heavy metals but also other salts, then you will have serious problems with an ion exchanger. It also depends on the scale of the plant: the larger the scale, the more it seems that the electrochemical processes are favoured. Another point is that it depends of course on the metal to be removed. If it is a metal for which regenerable ion exchange resins are available, then ion exchange is a good alternative. In some cases, however, the regeneration of the resins is not really possible. In such cases the process becomes quite costly.

J. Hoigné   •   Waste water from galvanic processes often contains very strong complex forming agents, in some cases even EDTA. Could you tell us how EDTA complexes behave upon cathodic reduction?

G. Kreysa • Even in the presence of very strong complex forming agents, there is an equilibrium concentration of free metal ions, and one of the advantages of three-dimensional electrodes is that they operate best at low concentrations. If there is a complexing agent present, then, due to the small local current density, high penetration depths are achieved. Another effect is that, at the cathode, not only free metal ions but also the complex itself can be reduced and hence destroyed. There are some processes running with EDTA-containing solutions as well as cyanide-containing solutions and in both cases they operate very well.

P.W. Prendiville • I believe you said that one of the applications was a 20 m^3/h waste water plant. Could you comment on what point in the waste water stream the process is introduced and could you also comment on how large a flow stream this process might be applied to?

G. Kreysa • I unfortunately cannot give you too many details about this specific process. It treats a copper containing waste water from a dyestuff production line. In the meantime, larger plants have been built, e.g. a 100 m^3/h plant in Italy for the removal of mercury from a waste water stream. There is no strict limitation with respect to higher throughputs, but there is only a low investment cost degression because the throughput of one m^2-cell is limited. The limitation depends, for example, on the conductivity of the electrolyte. If the conductivity is high, the throughput can be as high as two m^3/h in one m^2-cell, but there are cases where only 100 or 200 *l*/h can be put through such a cell. Here, a number of such cells have to be operated in parallel.

Advanced Ozone Generation

U. Kogelschatz
Brown Boveri, Baden, Switzerland

1. Introduction

Ozone was discovered in 1839 by C.F. Schönbein who studied the electrolytic decomposition of water. It took more than two decades of vehement scientific dispute before the constitution of this new substance was clearly identified as a three-atomic molecule containing only oxygen, namely O_3 (J.L. Soret 1865). About the same time in 1857, Werner von Siemens found out that ozone could also be generated in gas discharges and thus laid the foundations for modern industrial large-scale ozone production. This article will concentrate on the different aspects of ozone generation in gas discharges. Schönbein's idea of generating ozone by electrolysis has led to the development of modern electrochemical ozone generators for special applications (cf. Stucki and Baumann).

2. Ozone Properties and Applications

Ozone is an essentially colourless gas with a characteristic pungent odour. It is one of the strongest oxidants surpassed only by fluorine in its oxidizing power. This property makes it a potent germicide and viricide as well as a strong bleaching agent. Ozone also attacks the respiratory tract. The safe concentration is considered to be about 0.1 ppm, a value that is today occasionally surpassed in extreme smog situations. The ozone molecule is only moderately stable. In the absence of oxidizable substances, it decays to form the stable O_2 modification with a time constant of several days at room temperature. Elevated temperatures, UV radiation or the presence of catalysts can accelerate this decay drastically. The instability of the ozone molecule has two important consequences:

a)   Only special non-equilibrium gas discharges are suited for ozone generation, and

b)   Ozone can neither be shipped nor stored. It is always generated on site at a production rate that is controlled by the process.

The natural decay product of ozone is O_2. In the presence of organic substances CO_2 may also form. Neither of these substances is an environmental hazard. Thus, ozone has two definite advantages over other commercial oxidants: no transport or storage of potentially dangerous chemicals is involved in ozone applications and no objectionable by-products or residues are formed.

Historically, the major application of ozone has been in a process stage in the purification of drinking water (removal of taste, odour, colour, manganese, iron; reduction of turbidity; disinfection). The first large ozone plants were put into operation in Nice (1907) and in Petersburg (1910). Today, ozone is used in several thousand drinking water plants throughout the world. The 25 largest installations consume, on average, more than 1 MW for the ozone generators alone, indicative of the fact that ozone synthesis has become a major plasmachemical process.

In addition to its applications in the preparation of potable water, ozone is being increasingly used: in the treatment of waste water (disinfection, removal of colour, odour); in closed cooling circuits; in the treatment of industrial wastes (phenolic compounds, cyanide wastes); in bleaching processes (kaolin, textiles, wax, paper pulp); and in chemical synthesis (oxidation of oleic acid, production of hydroquinone, certain hormones and vitamins, camphor, perfumes). An oxidation step involving ozone is also being investigated as a means of NO_x removal from flue gases in power plants. In all these applications, ozone has one definite advantage over other oxidizing agents, namely, since only oxygen is added to the system, no undesirable residues or side products are formed. With growing interest in environmental protection, it is to be expected that ozone, because of its better environmental compatibility, will increasingly replace other chemical oxidants such as peroxides, permanganates, dichromates and chlorine or chlorine compounds. Improved equipment for reliable and cost-effective generation of large ozone quantities as well as more stringent pollution control legislation, will accelerate this replacement.

3. Discharge Physics and Reaction Kinetics

Certain chemical reactions can result in the formation of ozone. UV-radiation can also be employed to produce small amounts of ozone. Larger quantities of ozone, however, are exclusively produced using electrical discharges. The instability of the ozone molecule requires a very special non-equilibrium discharge, the "silent discharge" or dielectric-barrier discharge. Extensive reviews of the older literature can be found in references 1 and 2, a discussion of more recent contributions is given in reference 3. In this section, we summarize our present understanding of the discharge physics and chemistry of ozone formation in oxygen and in air. The more practical aspects of ozone generation will be treated in sections 4 and 5.

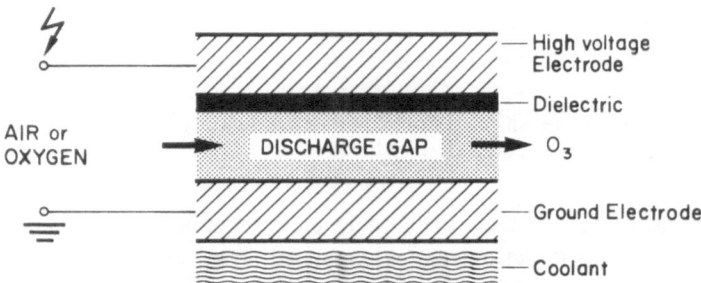

Fig. 1:   Electrode configuration of the silent or dielectric-barrier.

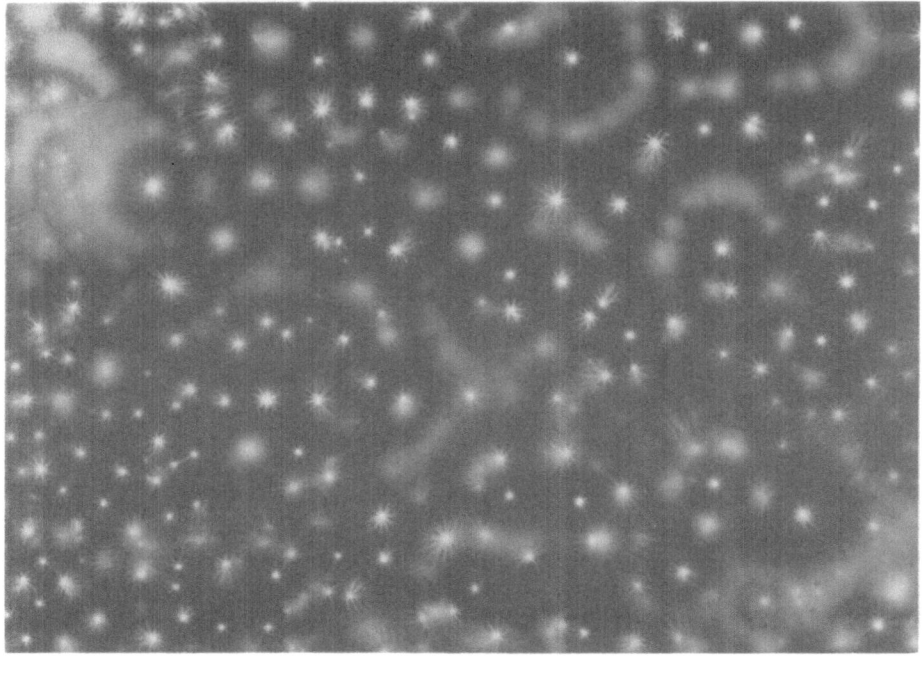

Fig. 2:   Photographic Lichtenberg figures showing the "footprints" of individual microdischarges for two different gap spacings.

3.1. Discharge Structure and Microdischarge Properties

Dielectric-barrier discharges are characterized by the presence of at least one dielectric layer in the current path between the discharge gap and the electrodes. Fig. 1 shows a schematic diagram of the electrode configuration of an ozonizer with one dielectric and active cooling of the ground electrode. The presence of the dielectric is essential in influencing the nature of the discharge. Since the current can pass the dielectric only in the form of a displacement current, these discharges can operate only in the AC mode. In air or oxygen near atmospheric pressure, the discharge is not homogeneous. In the discharge gap, the current is maintained by a large number of statistically distributed microdischarges of nanosecond duration. The most striking manifestations of these microdischarges are Lichtenberg figures which are used to visualize charge patterns on the dielectric surface. Fig. 2 shows two photographic Lichtenberg figures for two different gap spacings which were obtained by exposing photographic plates to the action of the discharge for about 1 ms. From the two Lichtenberg figures it is evident that the gap spacing has a strong influence on the strength of the microdischarges. Using the same technique, it can be shown that other parameters such as pressure, gas composition and humidity, as well as the nature and thickness of the dielectric and the feeding circuit can have an influence on the microdischarges. The properties of individual microdischarges were investigated with fast image intensifiers[4,5] and detailed current and charge measurements were made[6-8]. They can be summarized as follows.

Each microdischarge consists of an almost cylindrical current filament of about 100 μm radius in the discharge gap, which spreads into a surface discharge on the dielectric. Current densities up to 1000 A/cm^2 can be reached in the current filaments. Despite these high current densities, the transported charge (10^{-10} - 10^{-9}coulomb) and the energy density (about 10 mJ/cm^3) in the microdischarge channel are minute because of the extremely short duration of the current flow (2-5 ns). As a consequence, the gas temperature in the current filaments stays close to the average temperature in the discharge gap. The electron temperature, on the other hand, is determined by the electric field and reaches values of about 50,000 K, which corresponds to a mean electron energy of about 5 eV.

3.2. Initiation and Extinction of a Microdischarge

The breakdown voltage of a gas gap between two parallel electrodes is given by the Paschen curve. Paschen found out that the breakdown condition is a unique curve for each gas which depends only on the product pressure p times gap spacing d (at room temperature). Careful measurements have revealed that the same Paschen curve applies in the presence of a dielectric. Fig. 3 shows the Paschen curve for air in which the typical operating range of ozonizers is marked. Ozonizers operate near atmospheric pressure (typically between 0.5 and 3 bar) and use fairly narrow discharge gaps (typically 0.5 - 5 mm). Under these conditions, the electrical field required for breakdown is considerably higher than it would be in wider gaps. This is clearly shown by Fig. 4. It shows the reduced breakdown field E/n as a function of the product particle density n times gap spacing d for air and for oxygen. This presentation is equivalent to the Paschen curve but shows more clearly the physical parameters. The reduced electric field E/n determines the electron energy in the discharge, which is an important parameter influencing all electron collision phenomena. It is important to realize that, with narrow gaps, we can strongly influence the electric field at breakdown by varying the gap spacing or the particle density in the gap. Strictly speaking, these stationary breakdown curves apply only

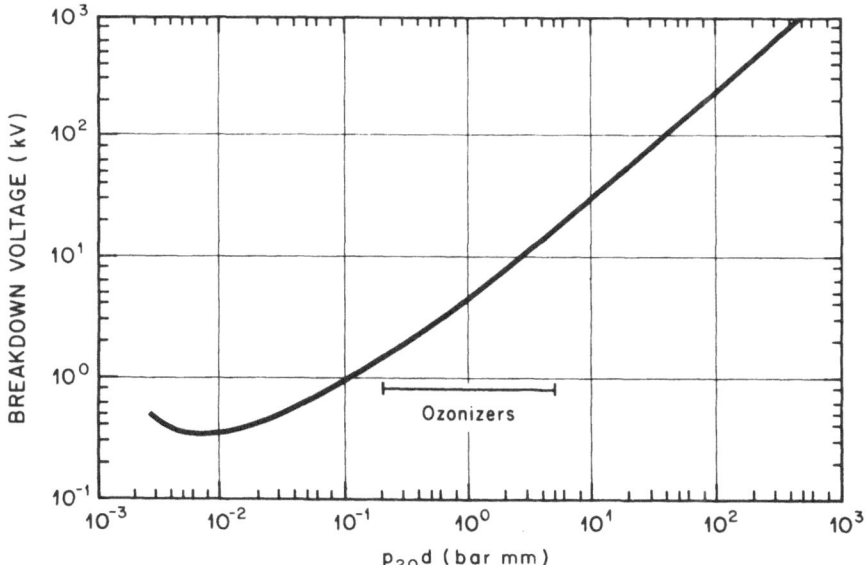

Fig. 3:   Paschen curve for air (at 20 °C) showing the breakdown voltage as a function of the product pressure times electrode spacing (after ref.9). The operating range of ozonizers is indicated.

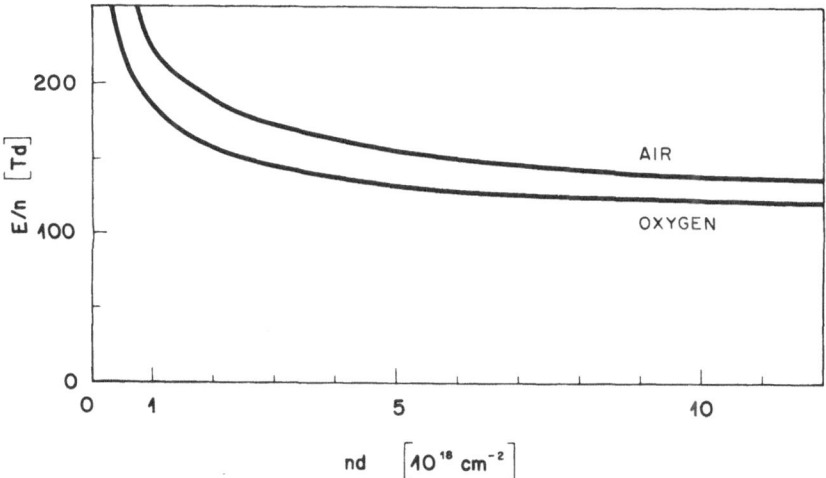

Fig. 4:   Reduced breakdown field as a function of the product particle density times electrode spacing. (1 Td = 10^{-17} Vcm^2, data from ref. 9 and 10).

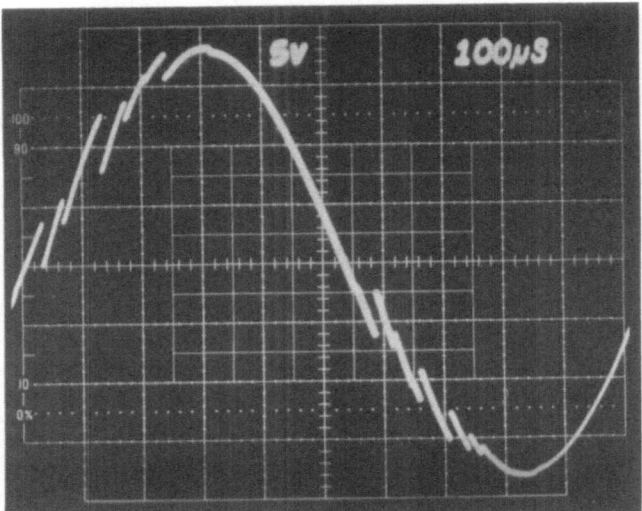

Fig. 5:   Ozonizer voltage showing strong pulsing.

for slowly varying applied voltages. Under certain conditions, namely extremely fast-rising voltages, or if large time delays occur, the electric field can overshoot the stationary breakdown condition. Considerable time delays (up to 1 ms) have been observed in narrow gap configurations.[11] Even moderately fast varying voltages, like those in modern medium frequency ozonizers (operating between 0.5 and 5 kHz) can, under these conditions, produce an overshooting of the breakdown condition. The oscillogram in Fig. 5 shows this effect quite clearly. The overshooting of the breakdown condition results in abrupt jumps of the ozonizer voltage. Each jump is caused by the simultaneous discharge of a large number of microdischarges (1,000 - 10,000), resulting in the discharge of a considerable fraction of the electrode area. This pulsing effect can be more or less pronounced depending on: the gas composition; the gap spacing; the pressure; and the electrode surfaces and the feeding circuit.

Which effect determines the duration of a microdischarge? As soon as the current flow is initiated in a microdischarge channel, charge accumulates in the area where the microdischarge hits the dielectric. As a result, the electric field in the gap is reduced at this location. When the field is reduced to such an extent that electron losses (mainly attachment) become more important than the electron production terms (ionization, detachment), the discharge is choked. From our current measurements of individual microdischarges, we know that this process is terminated within a few nanoseconds. It is well established[6,12] that the field does not collapse completely in the microdischarge channel. The next microdischarge at the same position can occur only after the breakdown condition is reestablished at this location. In the meantime, microdischarges occur at other positions. The dielectric thus has a dual function: it limits the charge that can flow through an individual microdischarge, and it spreads the microdischarges evenly over the entire electrode area.

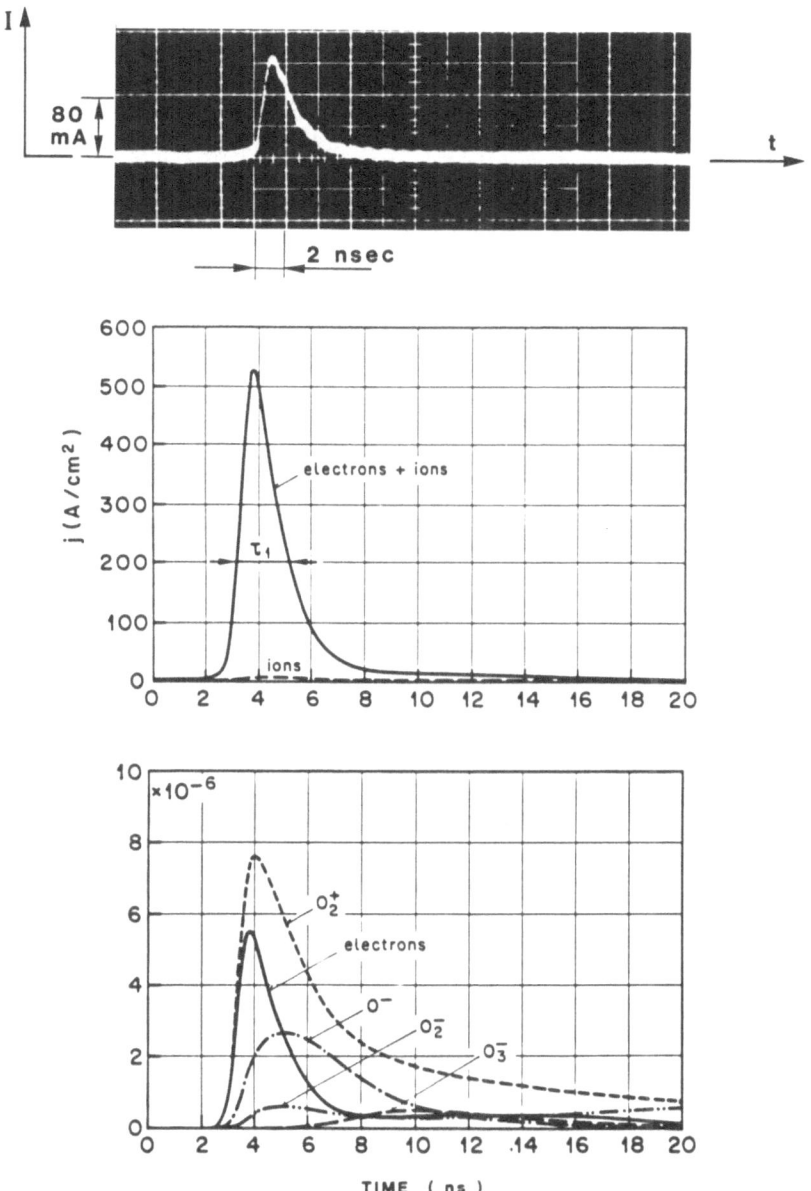

Fig. 6: Measured current pulse and numerical simulation of current density and normalized particle densities of charge carriers for a microdischarge in pure oxygen.

3.3. Discharge Modelling

Ozone is generated only in the microdischarges and not in the space inbetween. A volume element travelling through the discharge gap will be exposed to the action of a large number of microdischarges, typically several hundred microdischarges. A few authors have tried to model the physical and chemical changes in a microdischarge channel by solving rather complex reaction schemes[13-17]. When the breakdown voltage is reached, a very fast ionization process generates the charge carriers necessary for current flow. In pure oxygen, electrons and the ions O^+, O_2^+, O^-, O_2^-, O_3^- are the main charge carriers. Their relative importance depends on the strength of the microdischarge. In weak microdischarges, a considerable fraction of the energy is dissipated by ions. In stronger microdischarges, almost the entire discharge energy can be fed into the electrons. The strength of the microdischarge can be characterized by the energy deposition in the microdischarge channel or the relative oxygen atom concentration reached in the channel at the termination of the microdischarge. Our investigations[16] revealed that this quantity is several orders of magnitude higher than was assumed in previous publications[13,17,18]. Since the strength of the microdischarge has a strong influence on the efficiency of ozone generation, it is an important parameter for optimizing the performance of ozone generators. Fig. 6 shows a measured current pulse of a microdischarge together with a numerical simulation of electron and ion currents and the relative densities of the different charge carriers. It is interesting to note that, under these special conditions, the current is mainly an electron current, although the most abundant charge carriers are O_2^+ ions. This is due to the fact that the mobility of electrons is more than a factor 100 higher than that of the O_2^+ ions.

3.4. Ozone Formation in Oxygen

The electrons responsible for the current flow in a microdischarge can dissociate O_2 molecules via two processes with threshold energies of 6 and 8.4 eV, respectively:

$$e + O_2 \;\rightarrow\; e + O_2(A^3\Sigma_u^+) \;\rightarrow\; e + O(^3P) + O(^3P) \tag{1}$$

$$e + O_2 \;\rightarrow\; e + O_2(B^3\Sigma_u^-) \;\rightarrow\; e + O(^3P) + O(^1D). \tag{2}$$

The oxygen atoms can form ozone by the following reaction:

$$O + O_2 + M \;\rightarrow\; O_3 + M \tag{3}$$

in which M is a third collision partner, namely O_2, O_3, or in air also N_2. The two most important questions relating to ozone formation in pure oxygen are:

a) How much energy is necessary for the generation of an oxygen atom?

b) What fraction of the originally created oxygen atoms reacts to form O_3 molecules?

To answer the first question, detailed investigations into the different electronic processes revealed that, over a fairly wide E/n range, about 80% of the electron energy can be utilized for the dissociation process [19-21]. This situation is fairly unique in oxygen and is much less favourable in air. Fig. 7 shows the fraction of the electron energy going into different processes.

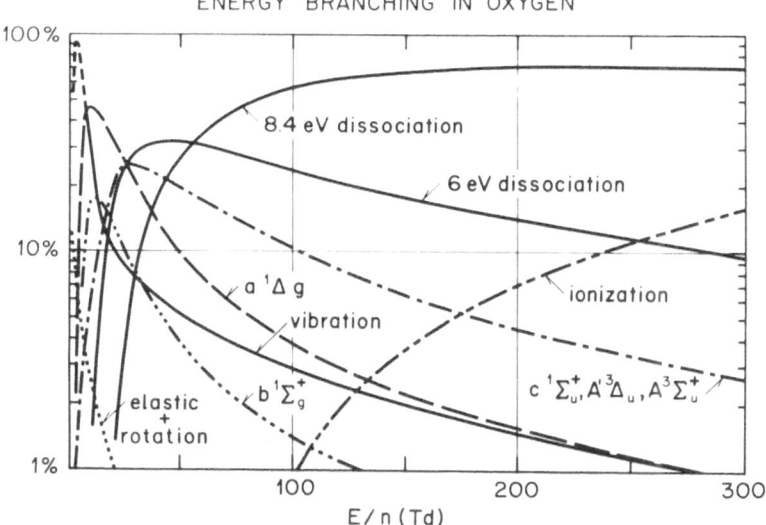

Fig. 7: Distribution of electron energy losses as a function of the reduced electric field.

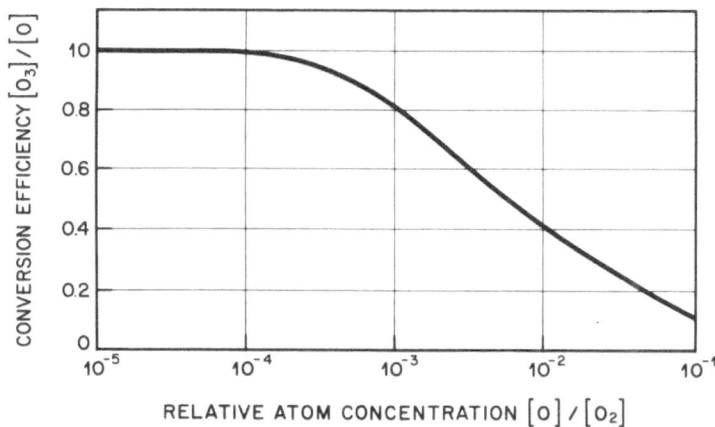

Fig. 8: Ozone yield per oxygen atom as a function of the relative atom concentration.

The fraction available for the dissociation of O_2 molecules is obtained by adding the 6 eV and the 8.4 eV process.

The maximum efficiency of ozone generation would be obtained if every O atom reacted according to reaction (3) to form an ozone molecule. Calculations[14,16] show that this is only possible for very weak microdischarges (Fig.8). In stronger microdischarges, additional chemical side reactions are significant:

$$O + O + M \rightarrow O_2 + M \tag{4}$$

$$O + O_3 \rightarrow 2O_2 \tag{5}$$

$$O + O_3^* \rightarrow 2O_2. \tag{6}$$

O_3^* stands for a transient excited ozone molecule which is the initial product of reaction (3). We may call these side reactions "chemical losses" because they either consume oxygen atoms which are then no longer available for ozone formation, or they destroy ozone molecules already formed.

We mentioned earlier that, in weak microdischarges, a considerable fraction of the discharge energy is dissipated by ions. Since ionic processes do not appreciably contribute to ozone formation, this situation has to be avoided. It turns out that, with respect to the efficiency of ozone formation, there exists an optimum discharge strength which is a compromise between avoiding excessive energy losses to ions and minimizing chemical losses due to chemical side reactions.

In an effort to optimize the discharge conditions we used rather complex reaction schemes, treating 70 reactions among 16 different particle species (ground states, excited states, charged particles) in pure oxygen[16]. Fig. 9 shows the concentrations of the major chemical species following a single microdischarge. One can see that the dissociation process is very fast and is completed at the end of the current pulse, while ozone formation is a much slower process which takes a few microseconds. These calculations predict a maximum efficiency of 400 g O_3/kWh if the energy dissipation by ions can be neglected. If we compare this number to that of the binding energy of ozone (143 kJ/mole or 0.82 kWh/kg), we can conclude that, at best, about one third of the discharge energy is channelled into ozone formation. The major part of the dissipated electrical energy is converted into heat and will have to be removed by

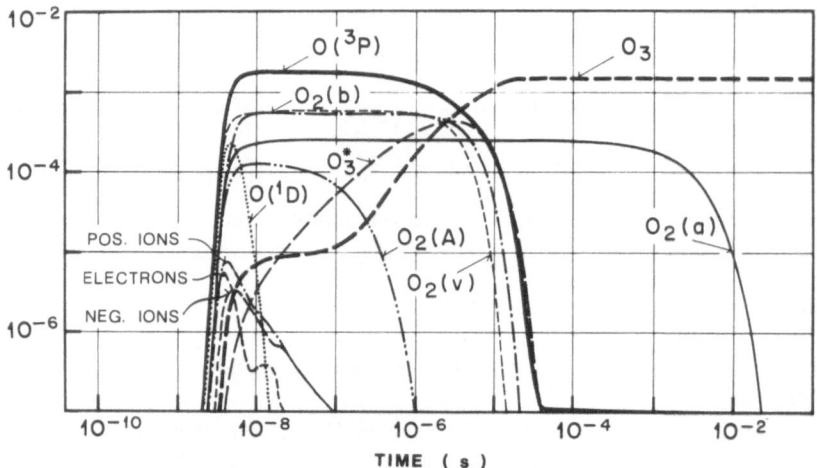

Fig. 9:   Evolution of different particle species following a microdischarge in pure oxygen.

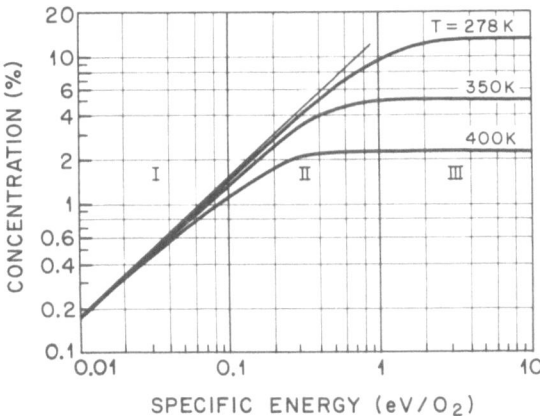

Fig. 10: Development of ozone concentration as a function of the specific energy for different temperatures.

the cooling circuit. In ozone discharge tubes, the highest experimental efficiency values at vanishing ozone concentration are about 250 g O_3/kWh. Higher values have been reported in very special laboratory experiments with pulsed discharges[22,23] or with glow discharges at very low temperatures[24].

We also calculated the rising ozone concentration due to the action of a large number of microdischarges. Fig. 10 shows initially a linear rise (region I), a slowing down (region II) and finally a constant plateau (region III), the saturation concentration. Once the saturation concentration is reached, each additional microdischarge will destroy as much ozone as it creates. Apparently this saturation concentration is very dependent on the temperature in the discharge gap.

3.5. Ozone Formation in Air

Ozone generation in air follows the same pattern as ozone generation in oxygen. We also observe individual microdischarges of slightly longer duration (about 5 ns) and slightly higher charge per microdischarge. The presence of other ionic species (N^+, N_2^+) and additional excitation and dissociation processes of nitrogen molecules add to the complexity of the system[25-29]. Reactions between oxygen and nitrogen species lead to the formation of a number of different nitrogen oxides. Fig. 11 shows the different particle species forming after a microdischarge in a mixture of 20% oxygen and 80% nitrogen. For this calculation, we used a reaction scheme treating 143 reactions between 30 reacting species. Our present understanding of ozone formation in air can be summarized as follows.

The different reactions leading to ozone formation have been identified. The nitrogen oxides NO, NO_2, NO_3, N_2O_5 and N_2O have been measured[25-30,34]. The major experimental trends can be predicted by computer simulation[27,34]. We will mention here only a few results. From the peaks of $O(^3P)$ and O_3 in Fig. 11, one can conclude that, again, ozone is formed from (ground state) oxygen atoms according to reaction (3). The abundance of oxygen atoms and also the ozone generating efficiency in air, however, is about twice as high as one could expect from

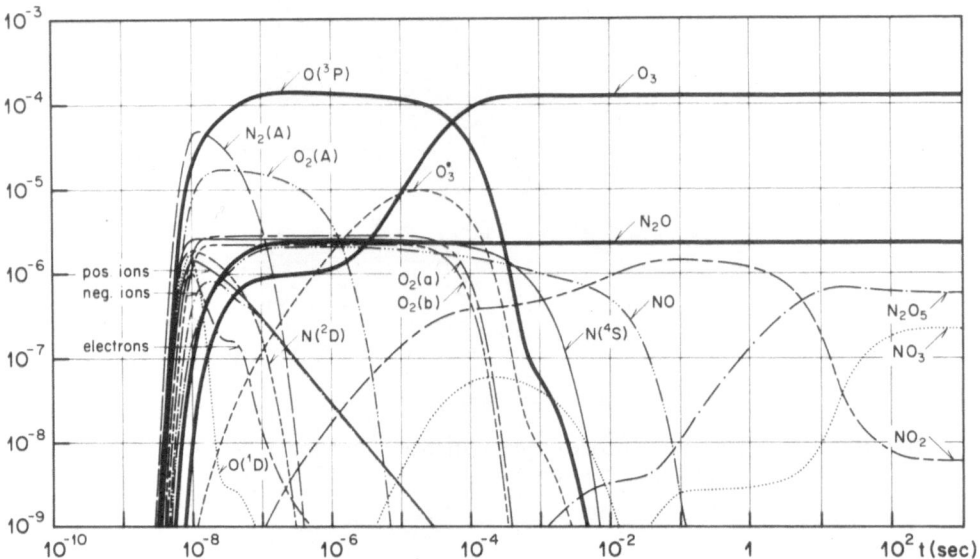

Fig. 11:   Evolution of different particle species following a microdischarge in
            "air" (20% O_2 + 80% N_2).

the oxygen content of air. Obviously there must be processes liberating oxygen atoms in
addition to direct electron-impact dissociation of O_2. A number of such processes could be
identified:

$$N_2(A^3\Sigma_u^+) + O_2 \rightarrow N_2O + O \tag{7}$$

$$N_2(A, B) + O_2 \rightarrow N_2 + 2O \tag{8}$$

$$N + O_2 \rightarrow NO + O \tag{9}$$

$$N + NO \rightarrow N_2 + O. \tag{10}$$

With these additional processes, the relative efficiency of ozone formation in different
N_2/O_2 mixtures could be predicted quite well by model calculations[17,26].

Under the normal operating conditions of commercial ozone generators, the nitrogen oxides
reach only about 1% of the ozone concentration[25,27,31-33]. At large specific energies, however,
they can inhibit ozone formation completely and destroy all previously generated ozone
(discharge poisoning, ozoneless mode[34]).

The two primary oxides formed within 100 ns after the microdischarge are nitrous oxide N_2O
(laughing gas) and nitric oxide NO. N_2O is a fairly inert species that is produced mainly by
excited nitrogen molecules via reaction (7) and does not react with the other nitrogen oxides.

NO is mainly formed in reaction (9) and, as the ozone concentration builds up, also in the
reaction

$$N + O_3 \rightarrow NO + O_2. \tag{11}$$

In the presence of ozone or oxygen atoms, NO is oxidized via NO_2 and NO_3 to the highest oxidation state N_2O_5. Thus, normally only N_2O_5 and N_2O can be detected in the output of air-fed ozonizers[33]. When discharge poisoning occurs, O_3 and N_2O_5 disappear and the lower oxides NO and NO_2 can be found in addition to N_2O.[34]

4. Laboratory Techniques and Macroscopic Parameters

4.1. Discharge Devices

Many different types of laboratory ozonizers have been described in the literature. Small parallel plate ozonizers have the advantage that the gap spacing can be varied in a simple way and that different electrode and dielectric materials can be studied. Plate ozonizers also furnished most of our present knowledge about microdischarge properties (Lichtenberg figures, image intensifier recordings, current and charge measurements). In our experience, however, it is very difficult to obtain high ozone concentrations in small plate ozonizers. One of the reasons is probably that the microdischarges close to the edge are different from those close to the center of the electrode area. It has also proved very difficult to obtain representative values for the ozone generating efficiency because the measurement of electric power presents a serious problem in such small devices.

Tubular ozonizers with an annular discharge gap have only minor edge effects. They can have an appreciable electrode area if they are long enough. We have made many measurements with a 1 m long discharge tube with a thermostat-controlled double-walled outer electrode made of stainless steel. A pyrex tube with an internal aluminum coating was mounted inside the steel tube to form an annular discharge gap (Fig. 12). The results obtained with such discharge tubes are very reproducible and, more importantly, are representative of those obtained in well-engineered large ozone generators containing several hundred discharge tubes.[36]

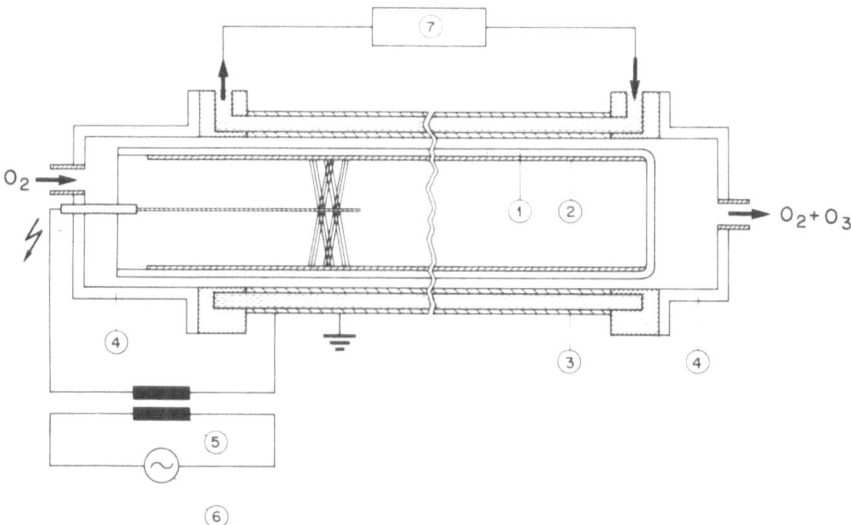

Fig. 12:   Cylindrical discharge tube with annular gap: (1) glass tube, (2) aluminum coating, (3) cooled stainless steel cylinder, (4) plexiglass end caps, (5) high voltage transformer, (6) AC power source, (7) thermostat.

U. Kogelschatz

4.2. Efficiency Curves

The performance of ozonizers is influenced by many parameters: gap spacing, pressure, temperature, gas composition, electrode and dielectric, and feeding circuit. For a given configuration, one of the strongest influences on the efficiency is that of the desired ozone concentration. Every ozonizer reaches its maximum efficiency at vanishing ozone concentration. The efficiency decreases monotonously with rising ozone concentration and drops to zero when the saturation concentration is reached. The maximum efficiency and the saturation concentration are the two most important parameters characterizing a certain discharge configuration. Fig. 13 shows a normalized efficiency curve for oxygen. The drawn curve is calculated on the basis of the action of a large number of microdischarges.[16] The circles are measured values. The shape of this curve is predicted quite accurately by the calculation. Both the maximum efficiency and the saturation concentration can be influenced by the experimental conditions and also by the design of the ozonizer. Efficiency curves have proved to be valuable tools for the optimization and comparison of different discharge devices.

4.3. Discharge Voltage

For the macroscopic description of dielectric-barrier discharges, the notion of the discharge voltage U_D is a useful quantity. Although the current flow through the discharge gap is maintained by a large number of short-lived microdischarges, we can define an average discharge voltage which is fairly constant during the active phase of the cycle. As long as the gap voltage U_g is smaller than U_D, there is no discharge activity and the device behaves like a series combination of two capacitances: the dielectric capacitance C_D and the gap capacitance C_g. As soon as the gap voltage reaches U_D, microdischarges are initiated and this activity continues

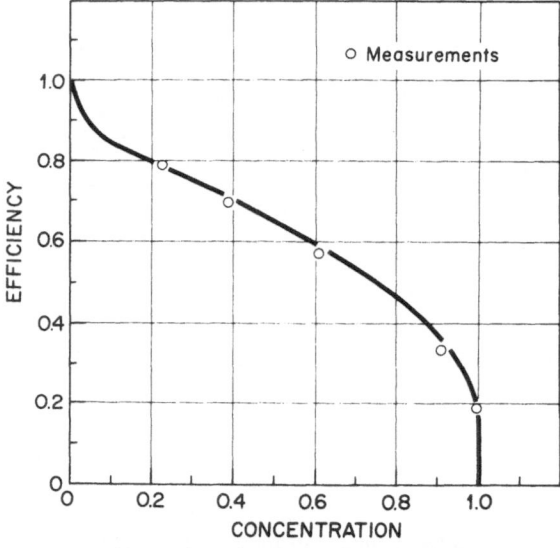

Fig. 13:   Normalized efficiency curve of an ozonizer operating on oxygen.

until the external voltage reaches its peak. During the active phase, the gap voltage remains approximately constant: $U_g = U_D$. There are two active phases per cycle of the fundamental frequency. The oscillograms of Fig. 14 show that the gap voltage is fairly constant during the active phases, even with strong pulsing action. These oscillograms were obtained by electronically subtracting the voltage across the dielectric, which can be obtained from an additional condenser in the circuit. The discharge voltage U_D has a value that lies between the voltage at the initiation of of microdischarge and the voltage after the microdischarge has been choked. Assuming no losses in the dielectric, we can derive an exact mathematical definition for U_D from the power relation

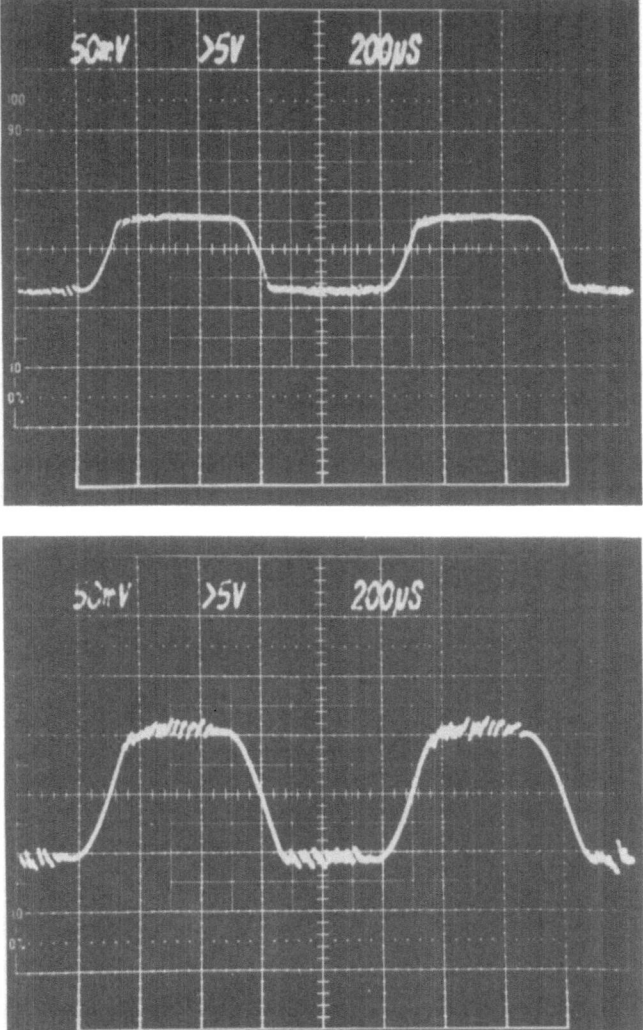

Fig. 14: Gap voltage of a silent discharge in the case of slight and strong pulsing action.

$$P \;=\; \frac{1}{T} \int_0^T U \, I \, dt \;=\; \frac{1}{\Delta T} U_D \int I \, dt \tag{12}$$

where ΔT is the time interval during which the discharge is on and the second integral is extended over the two active phases of the cycle. If we take one complete cycle of the fundamental frequency as the base for the definition of U_D, we also take care of any asymmetries that may exist between the positive and the negative half-waves. The discharge voltage is an average quantity that depends mainly on the gas composition, the particle density in the gap and the gap spacing. To a lesser extent, it also depends on the dielectric and on the metal electrode. It is one of the most important parameters used in the quantitative description of ozonizers. It can always be determined from oscillograms.

4.4. Power Formula

For the design of ozone installations, quantitative relations describing the power consumption are needed. Based on the finding that the discharge voltage U_D remains constant during the active phases of the cycle, fairly straightforward relations can be derived. Ozonizers are driven either by sinusoidal voltages or by thyristor-controlled frequency convertors that, in the simplest case, impress a square-wave current. For both cases, the ideal curves for external voltage, gap voltage and current are given in Fig. 15. The considerations below apply to both

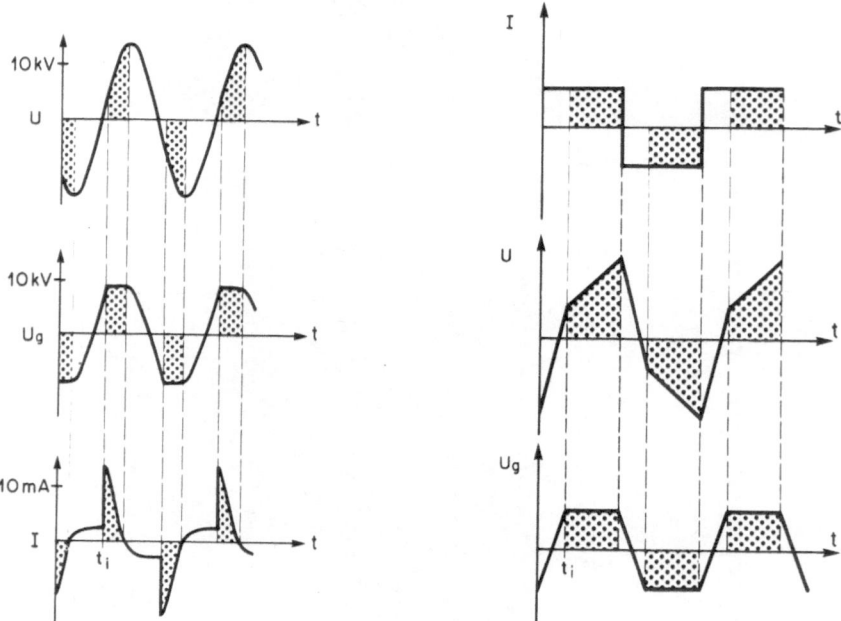

Fig. 15:   Ideal curves for external voltage, gap voltage and current for a silent discharge with sinusoidal feeding voltage and with impressed square-wave current. The time t_i indicates the moment of ignition, the dotted area the presence of the gas discharge.

cases. We assume that the dielectric barrier is an ideal capacitance and that $U_g = const. = U_D$ during the active phases of the cycle (discharge on). Assuming symmetry, we need only consider a half cycle. In general, the voltage change across a capacitance is related to the transported charge by:

$$\Delta Q = C \Delta U \tag{13}$$

According to Fig. 16, the charge transported between time t_0 and t_2 is

$$Q(t_0 \to t_2) = Q_{02} = C_D (2 \hat{U} - 2U_D) \tag{14}$$

because the total voltage changes from $-\hat{U}$ to $+\hat{U}$ ($\hat{U}$: peak value) and the gap voltage changes from $-U_D$ to $+U_D$. The charge necessary to change the voltage of the gap capacitance C_g (during the off phase) from $-U_D$ to $+U_D$ is

$$Q(t_0 \to t_1) = Q_{01} = C_g (2U_D). \tag{15}$$

From these two relations we can derive the charge transport during the active phase:

$$Q(t_1 \to t_2) = Q_{12} = Q_{02} - Q_{01} = (C_D [\hat{U} - U_D] - C_g U_D) \tag{16}$$

The discharge energy per half cycle is given by $U_D Q_{12}$ and the total power is obtained by multiplying by 2f (f: frequency):

$$P = 4 f C_D U_D (\hat{U} - \frac{C_D + C_g}{C_D} U_D) \tag{17}$$

or, with the abbreviations:

$$U_{Min} = (1 + \beta) U_D, \quad \beta = C_g / C_D, \tag{18}$$

$$P = 4 f C_D \frac{1}{1 + \beta} U_{Min} (\hat{U} - U_{Min}), \quad (U \ge U_{Min}). \tag{19}$$

This is the important power formula for dielectric-barrier discharges. It was first derived by T.C. Manley using a slightly different notation[37]. In our derivation, we did not make any assumptions about the shape of the external voltage, which means that relation (19) also holds for arbitrary voltage forms. In our notation, the power formula contains only measurable quantities: the frequency f, the minimum (external) voltage U_{Min} at which ignition occurs, and the peak voltage. A few points are worth mentioning:

a) For a given geometry (C_D,ß) the power depends only on f, $\hat{U}$, U_{Min}, (and not on the shape of the feeding voltage).

b) The discharge voltage U_D can immediately be calculated once U_{Min} is determined (relation 18).

c) For a fixed peak voltage and discharge voltage, the power is strictly proportional to the frequency and to the electrode area (C_D).

d) From Fig. 16 it is apparent that the external voltage at the moment of ignition (t_1) can be smaller than U_D. It can even be negative (the condition for ignition is $U_g = U_D$). The moment of extinction (t_2) is always at the extremum of the external voltage.

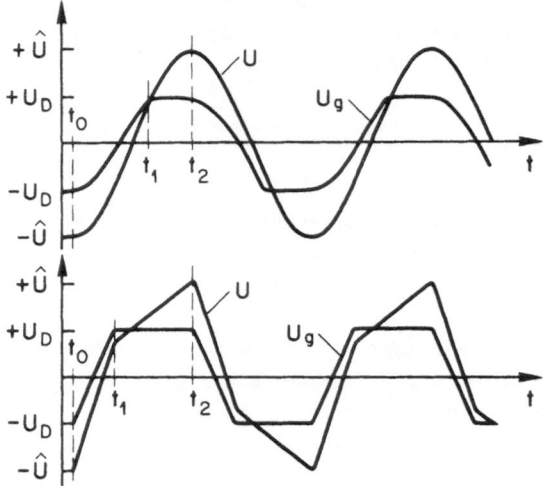

Fig. 16:   External voltage U and gap voltage U_g for
            sinusoidal feeding voltage and for square-wave-
            current.

Another simple expression for the power formula can be obtained when the average current

$$\overline{I} \;=\; \frac{1}{T} \int_0^T |I|\, dt \;=\; \frac{Q_{02}}{T/2} \;=\; 4f\, C_D\, (\hat{U} - U_D) \qquad (20)$$

is introduced[38]. In the general case, the average current $\overline{I}$ is not identical with the rms value I_{eff}. In the case of a square-wave current, however, these values are identical.

We can express the power as:

$$P \;=\; U_D\, (\overline{I} - \overline{I}_0) \qquad (21)$$

where $\overline{I}_0$ is the capacitive current at the moment the gap voltage reaches the value U_D:

$$\overline{I}_0 \;=\; 4f\, U_D\, C_g \;=\; 4f\, U_{Min}\, C_{tot} \quad \text{with} \quad C_{tot} \;=\; \frac{C_g C_D}{C_g + C_D}. \qquad (22)$$

Even large ozone installations follow these relations quite closely.

4.5 Power Measurement and U-Q Diagrams

The measurement of the discharge power can be a serious problem in small laboratory experiments, especially in the presence of strong pulsing action. In our experience, most power meters give misleading results and only expensive electronic power meters with true rms capability from DC to about 100 kHz can handle this task, provided that no phase error is introduced by the current and voltage measurement.

These difficulties can be avoided if the power is determined from a voltage-charge Lissajous figure (U-Q diagram). If a measuring capacitance C_M is introduced into the circuit and U_M is the voltage across this capacitance we can write:

$$I = C_M \frac{dU_M}{dt} \qquad \text{and} \qquad (23)$$

$$P = \frac{1}{T} \int_0^T U I \, dt = \frac{C_M}{T} \int_0^T U \frac{dU_M}{dt} dt = f C_M \oint U \, dU_M . \qquad (24)$$

When we put the ozonizer voltage U on the x-axis and the voltage U_M (as a measure of the charge Q) on the y-axis of an oscilloscope in the x-y-mode, we get a closed curve. The area inside the closed curve is proportional to the energy dissipated in the discharge during one cycle. The dielectric behaves like an ideal capacitance and does not noticeably dissipate power. This can easily be checked: as long as $U < U_{Min}$, the curve collapses to a line on the oscilloscope. This is also a check for phase errors of the high voltage divider.

For a dielectric-barrier discharge with constant discharge voltage during the active phase, this U-Q diagram is a parallelogram. With proper scaling (U in volts, Q in coulombs), the following relations hold (see Fig. 17):

$$\tan \gamma = 1/C_D, \quad \tan \alpha = 1/C_{tot} = \frac{C_D + C_g}{C_D \cdot C_g} \qquad (25)$$

From triangle 1 : $\quad x = U_{Min}/\sin \alpha$
From triangle 2 : $\quad y = x \sin(\alpha - \gamma)$ $\qquad\qquad$ (26)
From triangle 3 : $\quad z = (\hat{U} - U_{Min})/\sin \gamma .$

The area of the parallelogram is given by

$$A = 2y \cdot 2z = 4 \frac{\sin(\alpha - \gamma)}{\sin \alpha \sin \gamma} U_{Min} (\hat{U} - U_{Min}) \qquad (27)$$

$$= 4 (\frac{1}{\tan \gamma} - \frac{1}{\tan \alpha}) U_{Min} (\hat{U} - U_{Min}) \qquad (28)$$

$$= 4 C_D \frac{C_D}{C_D + C_g} U_{Min} (\hat{U} - U_{Min}) . \qquad (29)$$

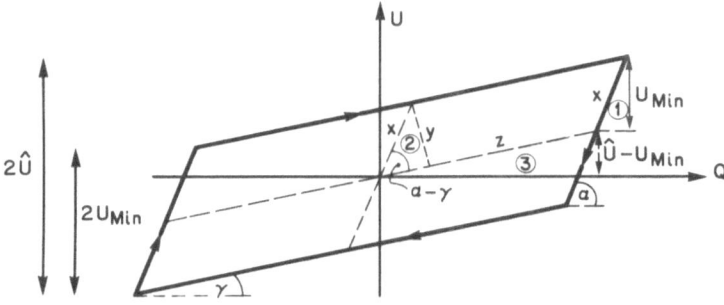

Fig. 17:   Idealized voltage-charge Lissajous figure (U-Q-diagram).

Taking into consideration that the area of the parallelogram is the discharge energy per cycle, we multiply by the frequency f and make use of the relation $\beta = C_g/C_D$:

$$P \;=\; 4\,f\,C_D\,\frac{1}{1+\beta}\,U_{Min}\,(\hat{U} - U_{Min})\,. \tag{19}$$

This relation is identical with the previously derived power formula (19), which can thus also be obtained on the basis of purely geometrical considerations. As a matter of fact, Manley[37] used similar geometrical considerations to derive the power formula. The oscillograms of Fig.18 demonstrate that the U-Q diagrams obtained from ozonizers are almost ideal parallelograms. Even in the case of strong pulsing action, they can be used in determining the discharge

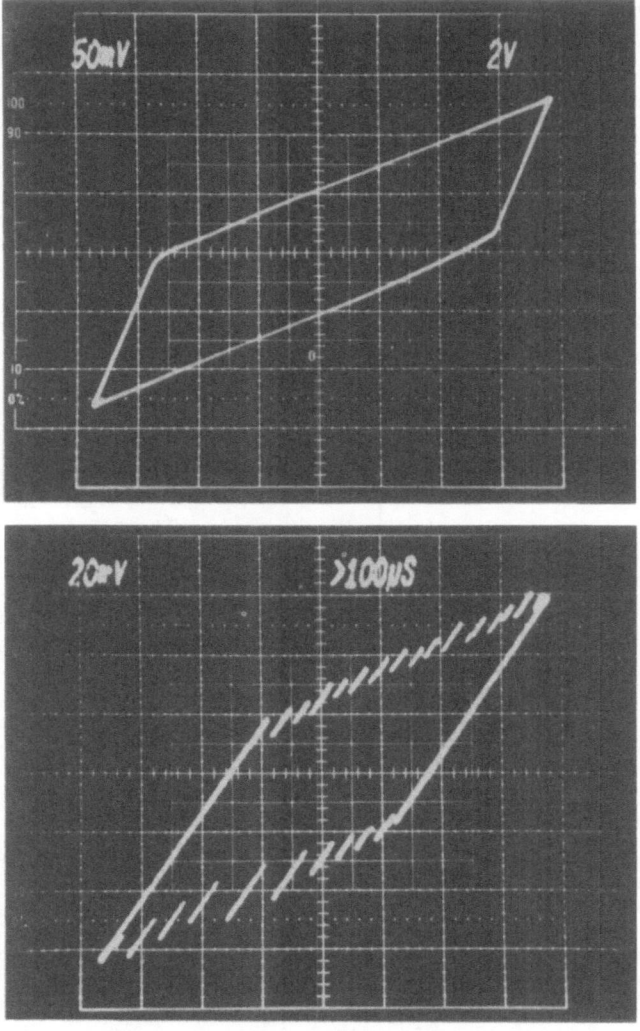

Fig. 18:   Lissajous figures (U-Q diagrams) in the case of negligible and strong pulsing action of the discharge.

parameters, especially the discharge power. Using modern transient digitizers and computer evaluation of the U-Q area, averaged over many cycles, this provides an alternative way of determining the discharge power in laboratory experiments.

4.6. The Power Parabola

A closer inspection of the power relation:

$$P = 4fC_D\frac{1}{1+\beta}U_{Min}(\hat{U} - U_{Min}) \tag{19}$$

shows that, for a given frequency, maximum power can be put into the discharge gap when:

$$\hat{U} = 2U_{Min} = 2(\beta+1)U_D. \tag{30}$$

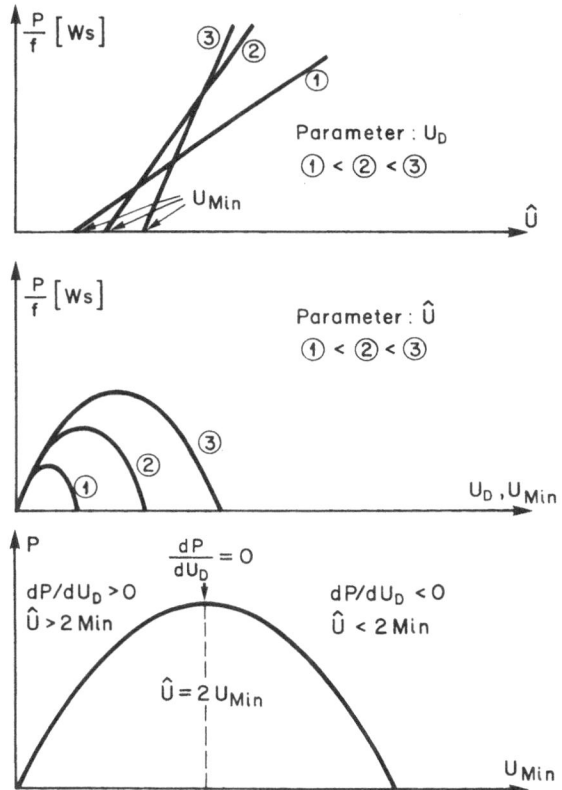

Fig. 19: Discharge power P as a function of peak voltage and discharge voltage U_D (f : frequency).

The relation between the power and the peak voltage is a linear one with the slope $4\,f\,C_D\,U_D$ and ignition at $\hat{U} = (1+\beta)\,U_D$. Fig. 19 shows some examples. The relation between the power and the discharge voltage is a quadratic one, describing the power parabola:

$$P = 4\,f\,C_D\,U_D\,(\hat{U} - (1+\beta)\,U_D). \tag{31}$$

The power parabola ranges from $U_D = 0$ to $U_D = \hat{U}/(1+\beta)$. U_D can, for example, be altered by varying the pressure in the discharge gap. U_D also changes slightly with ozone concentration. Since dP/dU_D is positive on the left side of the parabola and negative on the right side, this effect can be used to influence the power dissipation and consequently the gas temperature along the discharge gap. In fact, changing frequency and peak voltage we can obtain different positions on the power parabola for a given discharge voltage. Fig. 20 shows different examples with zero, positive and negative dP/dU_D all operating at a power density of $5\ \text{kW/m}^2$ of electrode area and an assumed discharge voltage of 6.8 kV.

4.7. The Power Factor

The power factor PF is an important electrotechnical quantity defined as:

$$PF = \frac{P}{U_{eff}\,I_{eff}}. \tag{32}$$

It determines how much reactive power has to be supplied by the power supply unit (PSU). Since ozonizers before ignition behave like a capacity load, the power factor can be considerably smaller than 1, typically between 0.2 and 0.8. It depends on the ratio of C_D/C_g, the ratio $\hat{U}/U_D$ and the voltage shape. In designing the PSU, it is essential to have exact values for the

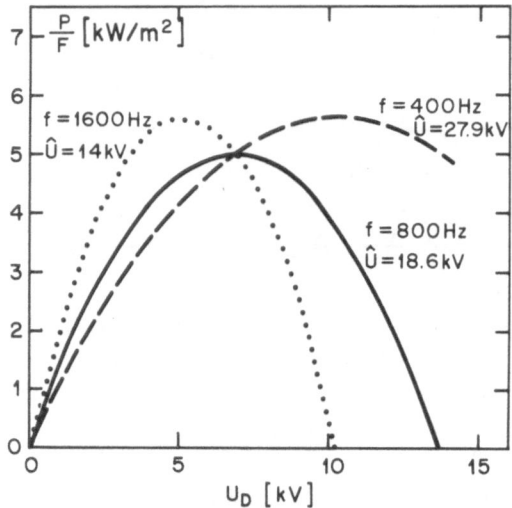

Fig. 20:   Demonstration of different operating points
            for the same power density (P/F = 5 kW/m²)
            and discharge voltage (U_D = 6.8 kV).

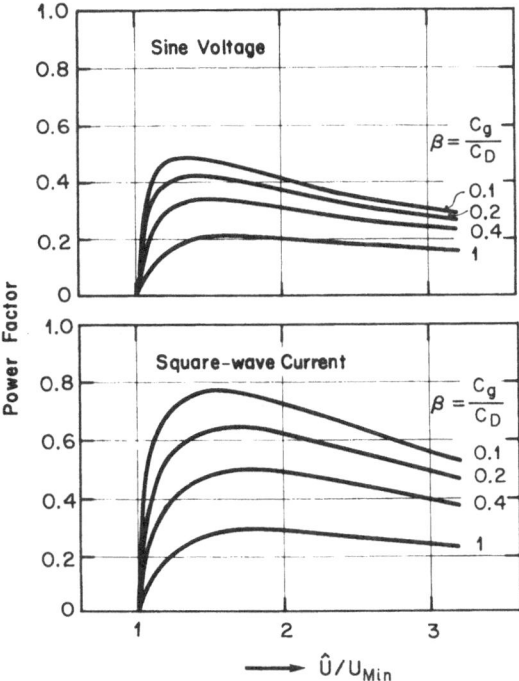

Fig. 21:   The power factor of a silent discharge configuration.

power factor under all operating conditions. The special cases of sinusoidal feeding voltage[39,40] and square-wave current[41] have been dealt with in the literature. Since the analytical expressions are rather lengthy, we give the results in graphical form (Fig. 21) and limit our discussion to a special case. The maximum of the power parabola is characterized by the condition $\hat{U} = 2\,U_{Min}$.

At this point of operation, the ignition of the discharge occurs when the voltage crosses the zero line (Fig. 22). Similarly, two corners of the U-Q diagram lie on the zero line. For this special condition, we obtain quite simple relations:

$$P \;=\; \frac{C_D}{\beta + 1}\, f\, \hat{U}^2 \tag{33}$$

<u>Sine Voltage</u>   <u>Square-Wave Current</u>

$$\left.\begin{aligned}
I_{eff} &= \pi\, f\, \hat{U}\; \frac{C_D}{1+\beta}\sqrt{1 + 2\,\beta(1+\beta)} & I_{eff} &= \bar{I} = 2f\, \hat{U}\; C_D \frac{1+2\beta}{1+\beta}\\[4pt]
U_{eff} &= \hat{U} / \sqrt{2} & U_{eff} &= \hat{U} / \sqrt{3}\\[4pt]
PF &= \frac{\sqrt{2}}{\pi}\, \frac{1}{\sqrt{1 + 2\beta(1+\beta)}} & PF &= \frac{\sqrt{3}}{2}\, \frac{1}{1 + 2\beta}
\end{aligned}\right\} \tag{34}$$

This point of operation can easily be reached by varying the voltage or current until ignition occurs at the zero line. At this point the discharge voltage can also be determined in a simple

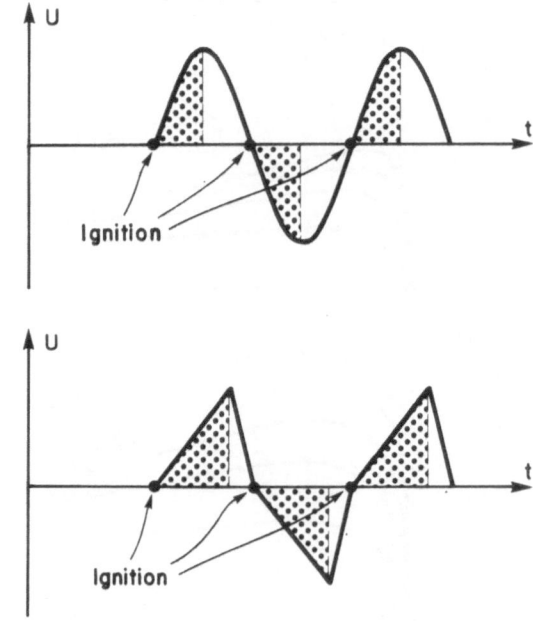

Fig. 22:   External voltage for a silent discharge operating at
the maximum of the power parabola in the case of
sinusoidal voltage and square-wave current (dotted
area: discharge is active).

way. For the sine voltage:

$$U_D = \frac{\hat{U}}{2(1+\beta)} = \frac{U_{eff}}{\sqrt{2}(1+\beta)} \tag{35}$$

and for the square wave current

$$U_D = \frac{\hat{U}}{2(1+\beta)} = \frac{\sqrt{3}\, U_{eff}}{2(1+\beta)} . \tag{36}$$

4.8. Ozone Measurements

In order to determine the efficiency of ozone generation, an exact measurement of the ozone production rate is needed as well as the measurement of the discharge power. This can be done by measuring the gas flow and the ozone concentration. The ozone concentration can be determined on-line by making use of the absorption properties of ozone (Fig.23). The strongest absorption is found in the ultraviolet part of the spectrum between 200 nm and 300 nm (Hartley band). Most ozone measurements are made at $\lambda = 253.7$ nm, which is close to the maximum absorption and is easily obtained using low-pressure mercury lamps. The absorption cross-section at this wavelength is known within 1% error[42,43]:

$$\sigma\,(253.65\ nm) = 1.14 \cdot 10^{-17}\ cm^2/molecule.$$

The ozone concentration can be calculated from the Lambert-Beer absorption law:

$$I = I_0 \exp(-n\sigma l), \tag{37}$$

where I and I_0 are the measured UV intensities passing the absorption cell with and without ozone present, n is the number of ozone molecules per cm^3 and l is the internal width of the

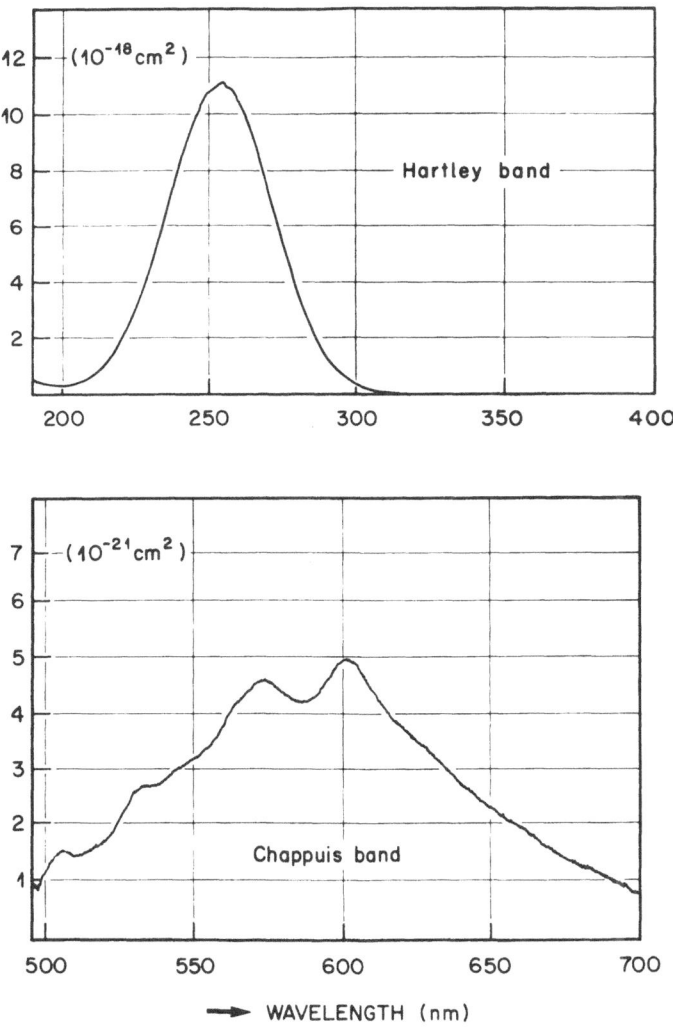

Fig. 23:   Absorption cross section of ozone in the ultraviolet and visible part of the spectrum (Resolution: 2nm).

absorption cell in cm (for experimental details, see the draft of the IOA European Standardisation Committee[44]).

The advantage of the UV absorption measurement is that it is instantaneous and can be used for process control without interference from other substances, such as the nitrogen oxides. Indeed, this method has been used with microsecond time resolution to measure the ozone formation time constants in air and in oxygen[26,45].

An alternative way of measuring ozone concentrations makes use of the much weaker absorption of ozone in the visible part of the spectrum (Chappuis band). The peak absorption is about 2000 times smaller than the UV absorption (Fig. 23) and requires correspondingly longer absorption paths. We obtained good results with a Helium-Neon laser ($\lambda = 632.8$ nm) and a 1 m absorption cell. Fowles and Wayne[46] describe a system with a 10 cm cell and a yellow LED as a light source.

It is also possible to make use of the strong infrared absorption bands of ozone at 4.7 μm and 9.5 μm, obtaining reproducible results.

The amount of ozone formed during a given time span can also be determined by chemical analysis[44,47]. Here the ozone containing process gas is bubbled through a buffered solution of potassium iodide (KI). Afterwards the solution is acidified and the liberated iodine in the solution is measured by titration with sodium thiosulphate (NaS_2O_3). Other oxidants of iodide, especially nitrogen oxides, will interfere with the measurement and give a positive bias. Under normal operating conditions of an ozonizer, however, the NO_x concentration will be about 100 times smaller than the ozone concentration. Thus, interference of NO_x will, under these conditions, be small.

4.9. Measurement of Nitrogen Oxides

During the past few years, the analysis of the different nitrogen oxide species has helped to establish the different reaction paths in air-fed ozonizers. Here we describe briefly how the concentrations of the different NO_x species were obtained. The sum parameter NO_x (= NO + NO_2 + NO_3 + $2N_2O_5$) can be monitored with commercial NO_x analysers[25]. In these instruments, a catalytic convertor reduces all higher nitrogen oxides to NO, which is then measured by the well-known chemiluminescence reaction between NO and added ozone:

$$NO + O_3 \quad \rightarrow \quad NO_2^* + O_2 \tag{38}$$

$$NO_2^* \quad \rightarrow \quad NO_2 + h\upsilon. \tag{39}$$

When the convertor is bypassed, the NO content is measured directly. Thus the fraction of NO and that of the higher nitrogen oxides can be obtained separately with these NO_x analysers. An identification of the higher oxides, however, is not possible with these instruments and N_2O completely escapes detection. If traces of water vapour are present, the NO_x output decreases due to the formation of HNO_3.

As mentioned previously (section 3.5), under normal operating conditions only N_2O_5 and N_2O, in addition to ozone, can be found in the output of ozonizers. This can clearly be shown with infrared absorption spectra covering the wavelength range of 3-16 μm. Fig. 24 shows the N_2O absorption band at 4.5 μm, the ozone bands at 4.7 and 9.5 μm and the three absorption bands of N_2O_5 at 5.8, 8 and 13.4 μm. These species have been measured over a wide range of operating parameters for modern high power ozonizers operating at medium frequencies[33].

When the specific energy (power/mass flow of feed gas) is raised to such an extent that dis-

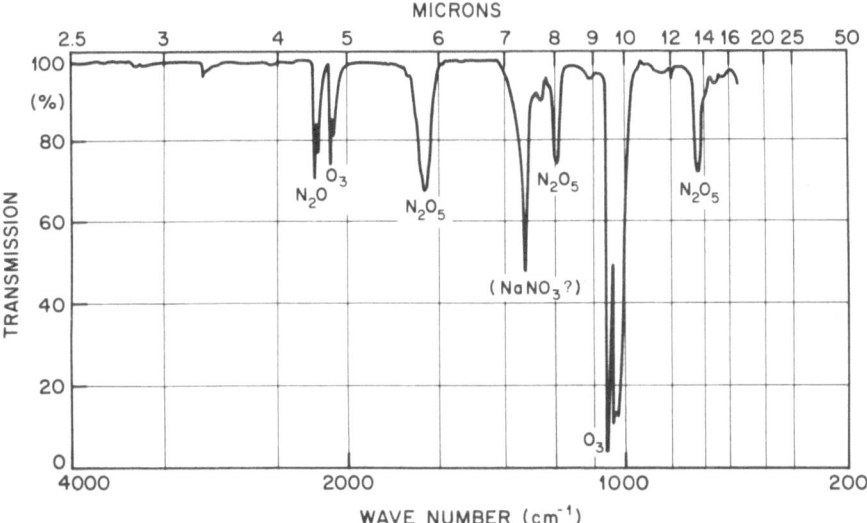

Fig. 24:   Infrared absorption spectrum of the gas mixture at the exit of an air-fed ozonizer (the absorption peak at 7.4 μm results from a deterioration of the NaCl cell windows).

charge poisoning occurs, the species O_3 and N_2O_5 disappear in the infrared spectrum and the NO_2 absorption band at 6.2 μm shows up besides the N_2O band at 4.5 μm. At this stage, NO also appears. It shows only very weak absorption in the infrared region (5 - 5.5 μm). It can be determined more easily from its characteristic UV absorption bands at $\lambda = 191, 195, 205, 215$ and 226 nm.

The nitrogen oxide species NO_3 can best be obtained under discharge conditions just before discharge poisoning occurs. The presence of NO_3 shows up in the visible part of the spectrum where two strong NO_3 absorption lines at 624 and 662 nm appear on top of the Chappuis absorption band of ozone. All these spectra can be found in reference 34.

An alternative way of determining different NO_x species is the use of a mass spectrometer immediately behind the exit of an ozone discharge[31,32].

5. Engineering Aspects

The better understanding of the ozone formation process in dielectric-barrier discharges has had considerable influence on the engineering of advanced ozone generators. If we compare the performance of modern ozone generators to the state of the art some years ago, we can identify three areas of improvement:

A.   higher power densities,

B.   higher ozone generating efficiencies, and

C.   higher ozone concentrations.

Traditional ozonizers operated at power line frequency (50 or 60 Hz) and reached typical power densities of 0.2 - 0.5 kW/m² of electrode area. They used fairly high operating voltages, large transformers and, when larger ozone quantities were required, huge installations. The use of medium frequency, thyristor-controlled power supplies operating between 0.5 and 5 kHz brought several advantages. The power density was increased by a factor of 5 to 10, which resulted in much smaller ozone generators. Also, the transformers decreased drastically in size. Since the thyristors are controlled by their gate pulses, the ozone generating process can be interrupted very fast in emergency by suppressing the gate pulses. In addition, process control by computer can be easily implemented.

For many of the older generators, ozone generating efficiencies of 40 - 100 g O₃/kWh were reported. These numbers depend very much on the preparation and the nature of the feed gas (air or oxygen) and, of course, on the desired ozone concentration. Modern ozone generators approach values of 100 g O₃/kWh in air and 250 g O₃/kWh in oxygen at low ozone concentrations.

As far as the maximum ozone concentration is concerned, considerable progress has also been made. It is possible, today, to build ozone generators that reach ozone concentrations up to 20% (by weight) in oxygen or 6% in air.

To reach these extreme values, the microdischarges have to be optimized with respect to certain properties, for example, high efficiency or high saturation concentration. The physical parameters in the microdischarges can be influenced by the gap spacing, the operating pressure, the composition of the feed gas, nature and thickness of the dielectric and, to some extent, also by the metal electrode. The temperature in the discharge gap is influenced by the power density and the design of the cooling circuit.

Ideally, an ozonizer should combine high power density, high ozone generating efficiency and high ozone concentration. But these are, in principle, conflicting requirements. Thus, it is not possible to reach optimum values for all of them with the same ozonizer and the same op-

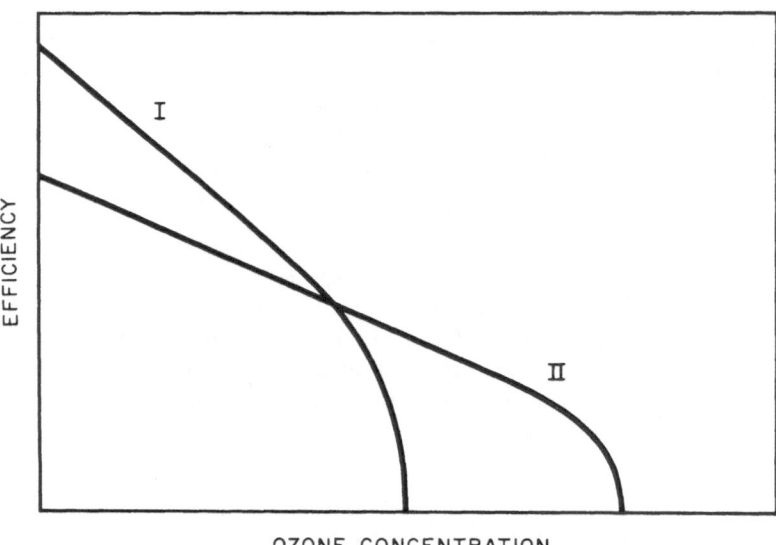

Fig. 25:   Efficiency curves for two different designs of ozone generators.

erating conditions. Due to our improved understanding of the physical processes in the microdischarges, we are in a position today, to tailor an ozone generator to the needs of the customer. The requirements for ozone applications in the chemical industry will, in general, be quite different from those in potable water preparation or waste water treatment. There may be yet different requirements for ozone used in bleaching kaolin or paper pulp.

The best ozonizer for a particular application has to be selected by its efficiency curve. Fig. 25 shows schematically the efficiency curves of two different ozonizers. The difference between these curves may be quite substantial. It is evident that for an application demanding high efficiencies at relatively low ozone concentrations, one would prefer an ozonizer according to curve I. On the other hand, when high ozone concentrations are required, we would certainly prefer curve II.

It should be pointed out that the ozone generator is only part of a complex system. For larger installations, a careful evaluation of the complete system will also include considerations about the composition and preparation of the feed gas and the handling of the process gas behind the ozonizer. We can distinguish between three different systems arranged in the order of increasing complexity:

a)   ozone generation from air,

b)   ozone generation from oxygen,

c)   ozone generation from oxygen with oxygen recycling.

For special applications, other gas mixtures, such as oxygen enriched air or oxygen/nitrogen/carbon dioxide mixtures, have also been proposed. The optimum configuration will, in general, depend on the availability and the price of oxygen. Only a larger system can afford to have a

Fig. 26:   Final assembly of Brown Boveri ozone generators (four out of five) for the Los Angeles Aqueduct Filtration Plant. Each generator has a nominal production rate of 37.5 kg ozone per hour at an ozone concentration of 6 weight percent.

cryogenic air separation stage to provide oxygen. Thus, for many of the smaller and medium size systems, operation on air will be the natural choice. It should be kept in mind, however, that air entering an ozone generator also costs something because it requires special preparation. Since the dew point has to be kept typically below -60 °C, the energy for compressing and drying the feed gas and the cost of the necessary installations both add to the overall cost. The total energy consumption of an air-fed system will be in the range of 18–20 kWh/kg ozone and typical ozone concentrations are 2 - 3%.

A good example of a large ozone installation employing oxygen is the new Los Angeles Aqueduct Filtration Plant. A fairly thorough evaluation for this installation is described in Ref. 48. The oxygen is produced on site. This plant has recently been put into operation. The five ozone generators (Fig. 26) can operate at concentrations up to 7%. Each of them can produce 37.5 kg ozone per hour. The most economic point of operation lies at 6% ozone concentration and requires in total 13.9 kWh/kg ozone which includes 4.6 kWh for the oxygen feed and 0.5 kWh for auxiliary power[36]. The possibility of having ozone without any traces of nitrogen oxides makes pure oxygen feed an attractive alternative for certain applications in smaller installations as well.

Very large ozone generators producing up to 150 kg O_3/hour each have been designed. Erni et al.[49] describe a large system with tonnage production of ozone for NO_x removal from flue gas of fossil fuel fired power plants. The ozone generating system consists of a cryogenic air separation unit, the ozone generators and an oxygen/ozone separation stage. In this separation stage, the remaining oxygen is separated from the ozone by pressure swing adsorption in silica gel columns and is recirculated to the entrance of the ozone generator. Desorption of the ozone is accomplished at a lower pressure with a stream of nitrogen from the air liquefaction stage. The ozone concentration in the final product gas stream can reach 6 to 10 wt% in the nitrogen carrier gas. The overall energy consumption is 10 -12 kWh/kg ozone.

In the first larger ozone installation for water treatment (Nice 1907), the ozone generators alone required 25 - 28 kWh/kg ozone generated and the total energy consumption was about 140 kWh/kg ozone applied to the water[50]. If we compare these figures to those of today's advanced ozone generating systems, we can safely conclude that the detailed investigations into the discharge physics and reaction kinetics of ozone formation as well as advanced engineering concepts and system optimization have resulted in considerable improvements in the performance and reliability of modern ozone generating systems.

Acknowledgement: Thanks are due to many colleagues who, at some time or other, contributed to this research effort: B. Adam, P. Baessler, R. Duerst, B. Eliasson, P. Erni, M. Fischer, U. Gloor, M. Hirth, R. Kirchhofer, W. Knapp, G. Mechtersheimer, J. Müller, D. Suter and H.J. Wiesmann.

References

1   Lunt, R.W., The mechanism of ozone formation in electrical discharges. Adv. Chemistry Series **21** (1959) 286-303.

2   "Ozon: Bildung und Zerfall auf elektrischem Wege" in *Gmelin Handbuch der anorg. Chemie, Sauerstoff, Syst. Nr. 3*, (Verlag Chemie GmbH, Weinheim) 1960, pp. 1038-1077.

3   Kogelschatz, U., Ozone synthesis in gas discharges. Proc. XVI Int. Conf. on Phenomena in Ionized Gases, Düsseldorf 1983, Invited papers, pp. 240-250.

4   Tanaka, M., Yagi, S., and Tabata, N., The observation of silent discharge by image intensifier. Trans. IEE of Japan **98 A** (1978) 57-62.

5   Heuser, C., Zur Ozonerzeugung in elektrischen Gasentladungen (Ph.D.-Thesis, RWTH Aachen) 1985.

6 Hirth, M., Teilprozesse bei der Ozonerzeugung mittels stiller elektrischer Entladung. I. Die elektrische Entladung im Ozonisator. II. Die Ozon- und Stickoxydbildung im Ozonisator. Beitr. Plasmaphys. **20** (1981) 1-27.

7 Hirth, M., Kogelschatz, U. and Eliasson, B., The structure of the microdischarges in ozonizers and their influence on the reaction kinetics. Proc. 6th Int. Symp. on Plasma Chemistry, Montreal 1983, pp. 663-668.

8 Mechtersheimer, G., Eliasson, B. and Kogelschatz, U., Polarity resolved measurements of the ozone production efficiency. Proc. XVIII Int. Conf. on Phenomena in Ionized Gases, Swansea 1987, pp. 522-523.

9 Dakin, T. W., Luxa, G., Oppermann, G., Vigreux, J., Wind, G. and Winkelnkemper, H., Breakdown of gases in uniform fields, Paschen curves for nitrogen, air and sulfur hexafluoride. Electra **32** (1974) 61-68.

10 Blair, D.T.A. and Whittington, H.W., Ionization and breakdown in oxygen. J. Phys. D: Appl. Phys. **8** (1975) 405-415.

11 Shibuya, Y., Breakdown time lag of short gaps in various gases. Proc. 3rd Int. Conf. on Gas Discharges, London 1974, pp. 132-135 (IEE Conf. Publ. No. 118).

12 Bertein, H., Charges on insulators generated by breakdown of gas. J. Phys. D: Appl. Phys. **6** (1973) 1910-1916.

13 Gibalov, V.I., Samoilovich, V.G. and Filippov, Yu.V., Physical chemistry of the electrosynthesis of ozone. The results of numerical experiments. Russ. J. Phys. Chem. **55** (1981) 471-479.

14 Eliasson, B., Hirth, M. and Kogelschatz, U., Ozone formation in dielectric-barrier discharges in oxygen. Proc. 7th Int. Symp. on Plasma Chemistry, Eindhoven 1985, pp. 339-344.

15 Peyrous, R., Numerical simulation of the production of neutral gaseous species created by electrical discharges in moist oxygen or air. Proc. 8th Int. Conf. on Gas Discharges and their Applications, Oxford 1985, pp. 489-492.

16 Eliasson, B., Hirth, M. and Kogelschatz, U., Ozone synthesis from oxygen in dielectric-barrier discharges. J. Phys. D: Appl. Phys. **20** (1987) 1421-1437.

17 Yoshida, K. and Tagashira, H., Computer simulation of ozone electrosynthesis in an N_2/O_2 mixture-fed ozonizer. Memoirs of the Kitami Inst. of Technol. **18**, 1 (1986) 11-20.

18 Yagi, S. and Tanaka, M., Mechanism of ozone generation in air-fed ozonizers. J. Phys. D: Appl. Phys. **12** (1979) 1509-1520.

19 Fournier, G., Bonnet, J. and Pigache, D., Comparaison des propriétés macroscopiques des électrons soumis à l'action d'un champ électrique dans l'air sec et dans l'oxygène pur. C.R. Acad. Sc. Paris **290** (1980) B179-B182.

20 Penkin, N.P., Smirnov, V.V. and Tsygir, O.D., Investigation of the electrokinetic properties and of the dissociation of O_2 molecules in an oxygen discharge. Sov. Phys. Techn. Phys. **27** (1982) 945-949.

21 Eliasson, B. and Kogelschatz, U., Electron impact dissociation in oxygen. J. Phys. B: At. Mol. Phys. **19** (1986) 1241-1247.

22 Salge, J., Kaerner, H., Labrenz, M., Scheibe K. and Braumann, P., Characteristics of ozonizers supplied by fast rising voltages. Proc. 6th Int. Conf. on Gas Discharges and their Applications, Edinburgh 1980, pp. 94-97 (IEE Conf. Publ. No. 189).

23 Yamabe, C., Akiyama, H. and Horii, K., The improvement of ozone yield by the high frequency corona discharge superposed on the pre-ionization. Proc. 7th Int. Symp on Plasma Chemistry, Eindhoven 1985, pp. 327-332.

24 Masuda, S. and Koizumi, S., Production of ozone at cryogenic temperatures by glow discharge and high frequency surface discharge. Proc. 8th Int. Symp. on Plasma Chemistry, Tokyo 1987, pp. 769-774.

25 Yagi, S., Tanaka, M. and Tabata, N., Generation of NO_x in ozonizers. Trans. IEE of Japan **99** (1979) 41-48.

26 Eliasson, B., Kogelschatz, U. and Baessler, P., Dissociation of O_2 in N_2/O_2 mixtures. J. Phys. B: At. Mol. Phys. **17** (1984) L797-L801.

27 Eliasson, B. and Kogelschatz, U., N_2O formation in ozonizers. J. Chim. Phys. **83** (1986) 279-282.

28 Samoilovich, V.G. and Gibalov, V.I., Kinetics of the synthesis of ozone and nitrogen oxides in a barrier discharge. Russ. J. Phys. Chem. **60** (1986) 1107-1116.

29 Okazaki, S., Kubo, S., Niwa, H., Kogoma, M., Sugimitsu, H., Moriwaki, T. and Inomata T., Ozone formation from the reactions of O_2-activated N_2 molecules and new type ozonizer with fine wire electrode. Proc. of the Symp. on Ozone + Ultra Violet Water Treatment, Aquatech, Amsterdam, 1986, pp. A.4.1-A.4.16.

30 Becke, Ch. and Maier, D., Nebenprodukte bei der Ozonerzeugung. Tagungsband Wasser Berlin '81 .

5. Ozon-Weltkongress, Berlin 1981, pp. 860-874.

31 Gibalov, V.I., Samoilovich, V.G. and Wronski, M., Electrosynthesis of nitrogen oxides and ozone in an ozonizer. Proc. 7th Int. Symp. on Plasma Chemistry, Eindhoven 1985, pp. 401-406.

32 Samoilovich, V.G., Gibalov, V.I. and Wronski, M., The mechanism of nitrogen oxides and ozone electrosynthesis in ozonizer. Proc. XVII Int. Conf. on Phenomena in Ionized Gases, Budapest 1987, pp. 325-326.

33 Kogelschatz, U. and Baessler, P., Determination of nitrous oxide and dinitrogen pentoxid concentrations in the output of air-fed ozone generators of high power density. Ozone Sci. Eng. **9**, 3 (1987) 195-206.

34 Eliasson, B. and Kogelschatz, U., Nitrogen oxide formation in ozonizers. Proc. 8th Int. Symp. on Plasma Chemistry, Tokyo 1987, pp. 736-741.

35 Kubo, S., Kogoma, M., Inomata, T., Sugimitsu, H., Moriwaki, T. and Okazaki, S., Formation de l'ozone par effet couronne sur une electrode très fine. J. Chim. Phys. **84**, 1 (1987) 87-91.

36 Erni, P., Fischer, M. and Liechti, P., Large scale ozone production from oxygen: First experience with the Los Angeles Aqueduct Filtration Plant. Proc. 8th Ozone World Congress, Zürich 1987, pp. A19-A27.

37 Manley, T.C., The electrical characteristics of the ozone discharge. Trans. Electrochem. Soc. **84** (1943) 83-96.

38 Samoilovich, V.G. and Filippov, Yu.V., Electrical theory of ozonizers VIII. Effect of frequency on the electrical characteristics of ozonizers. Russ. J. Phys. Chem. **35** (1961) 94-96.

39 Filippov, Yu.V. and Emel'yanov, Yu.M., The electrical theory of ozonizers V. The power factor of ozonizers. Russ. J. Phys. Chem. **33** (1959) 155-159.

40 Faes, Y., Ozoneurs: leurs théorie et application des techniques nouvelles à semi-conducteurs à leur alimentation. Rev. Gén. d'Electricité **84**, 1 (1975) 13-23.

41 Tabata, N., High-frequency ozonizer driven by current impressed type inverter. Proc. 2nd Int. Symp. on Ozone Technology, Montreal 1975, pp. 120-131.

42 Mauersberger, K., Barnes, J., Hanson, D. and Morton, J., Measurement of the ozone absorption cross-section at the 253.7 nm mercury line. Geophys. Res. Lett. **13**, 7 (1986) 671-673.

43 Molina, L.T. and Molina, M.J., Absolute absorption cross sections of ozone in the 185 to 350 nm wavelength range. J. Geophys. Res. **91**, D 13 (1986) 14.501-14.508.

44 Masschelein, W.J., Methods for the control of ozone in a process gas. Ozone News **15**, 4 (1987) 8-11 (Draft of the IOA European Standardisation Committee).

45 Sugimitsu, H. and Okazaki, S., Measurement of the rate of ozone formation in an ozonizer. J. Chim. Phys. **79** (1982) 655-660.

46 Fowles, M. and Wayne, R.P., Ozone monitor using an LED source. J. Phys. E: Sci. Instrum. **14** (1981) 1143-1145.

47 Maier, D. and Kurzmann, G.E., "Determination of high ozone concentrations in air" in *Analytical Aspects of Ozone Treatment of Water and Waste Water*, by R.G. Rice, L.J. Bollyky and W.L. Lacy, eds. (Lewis Publishers, Inc., Chelsea, MI, USA) 1986, pp. 271-292.

48 Monk, R.O.G., Yoshimura, R.Y., Hoover, M.G. and Lo, S.H., Prepurchasing ozone equipment. J. AWWA **77**, 8 (1985) 49-54.

49 Erni, P. Fischer, M. and Klein, H.-P., Tonnage production of ozone for NO_x-removal from flue-gas. Proc. 7th Ozone World Congress, Tokyo 1985, pp. 79-84.

50 Terrade, G., Drinking water treatment with ozone was born in Nice 80 years ago. Ozone News **14**, 5 (1986) 13-16.

Discussion

Chairman: Prof. M. Mirbach
Brown Boveri, Baden, Switzerland

M. Mirbach • On one of the slides (Fig. 10) you show that the saturation concentration of ozone was at 10%. Is this the highest you can get?

U. Kogelschatz • No, in oxygen we can get up to 20% wt., in air we can get up to 6% wt.

M. Mirbach • And from which concentration on is it explosive?

U. Kogelschatz • Well, if you look through the literature they say 14%, but we have never seen an explosion. We have been up to 20%.

J. Hoigné • At the IOA-Conference last week we heard that in Sipplingen (Lake of Constance) a stripping tower to remove N_2O_5 from the ozone gas was installed. Do you think that is also required for a new ozonator or is this only needed when the operating system is wrong?

U. Kogelschatz • Well, we have measured the N_2O_5 concentrations under normal operating conditions[33] and we find that with a typical dosage of 1 mg of ozone per liter of drinking water about 20 microgram of N_2O_5 are dissolved as HNO_3. I think this concentration is not a problem in drinking water.

P. Francis • I think in your introduction you mentioned a reduction of ten in the electrode area but you also mentioned the factor of two, I think, in energy. This was comparing your high frequency ozonizer with the 50 Hz ozonizer. Is it that much improvement in the efficiency?

U. Kogelschatz • Well, yes. The savings in the area are mainly given by the increase of the frequency from 50 to 600 or 800 Hz. The efficiency improvements are not only due to the feeding circuit, it's mainly better engineering of the ozone generator. The numbers refer to a comparison of equipment that was built 10 years ago and the best modern equipment available today for larger installations.

P. Francis • The other question is: which oxides of nitrogen do you think are important in reducing the efficiency of ozone formation when you have damp air as a feed gas?

U. Kogelschatz • I didn't go into that. In the output of ozonizers we find only two of these oxides: N_2O_5 and N_2O. The lower nitrogen oxides NO and NO_2 are not found in the output of the ozonizer because they are oxidized rapidly by ozone to the highest state of oxidation, which is the N_2O_5. In the case of damp air we find HNO_3 instead of N_2O_5. We also observe a drastic change in the strength of the microdischarges which may also be a reason for the reduction in efficiency.

E. v. Naerssen • There are some developments with respect to the size of the tubes and the material of the dielectric. Could you comment on these, with respect to their relevance to size and power consumption?

U. Kogelschatz •   The size of the tubes doesn't have much of an effect on the efficiency. It is a matter of engineering whether you prefer small tubes or larger tubes. As a matter of fact, you can get more electrode surface into a given volume using smaller tubes. But as long as you keep constant the type and the thickness of the glass, then you will not see large effects of the tube diameter. Changing the dielectric, however, has a very large influence on the efficiency. In most cases alternative dielectrics are worse than glass. Glass is still a very good dielectric material. Some ozonizers with ceramic materials are now on the market. We have done laboratory experiments with ceramics, but so far I see no indication to use ceramics instead of glass.

H.P. Klein •   Ceramic dielectrics are exclusively used in small units with very high frequencies, i.e. about 10,000 Hz for ozone production rates of a few grams – up to a few 100 g perhaps. It is an expensive way of producing ozone, because the dielectric is expensive and the dimensions of the available ceramic materials are not very large.

G. Kreysa •   I was very impressed by the clear evidence you gave for the fact that nitrogen oxides will destroy ozone. I know from the literature in connection with the ozone hole in the stratosphere above the Antarctic that several authors agree that nitrogen oxides alone cannot be responsible for ozone depletion, because these reactions involve the formation of radicals which can regenerate ozone. These authors claim that the ozone destruction in the atmosphere is mainly due to the chlorinated hydrocarbons. What is your opinion on this? I know that we are talking about quite different concentration ranges, but if you made model calculations on this, perhaps you can give us your opinion on this.

U. Kogelschatz •   O.k. let me repeat it. It's a little more complicated. I didn't say that the nitrogen oxides destroy ozone. They only use up ozone when they are oxidized in a stoichiometric reaction using, I think, 1.5 ozone molecules for one NO. I said that nitrogen oxides interfere with an intermediate step at the level of the atomic oxygen and that they can prevent the ozone formation in an ozonizer. This interference is a typical situation in a gas discharge. In atmospheric physics I agree that it's probably not the nitrogen oxides that are important. It may be – it's not established yet – that in fact the chlorinated and fluorinated hydrocarbons are responsible. It's probably not a pure gas phase reaction; droplets and condensation may also play a role in these phenomena.

The Chemistry of Ozone in Water

J. Hoigné
EAWAG, Dübendorf, Switzerland

Abstract

From lists of published rate constants for the direct reactions of aqueous ozone with substrates we can conclude: if, for example, 0.5 mg/l ozone is present, the following types of compounds react within less than 10 seconds: sulfite, nitrite, olefinic aliphatic hydrocarbons, phenols, polyaromatic hydrocarbons, organic amines and sulfides. Much slower to react are some chlorosubstituted olefins (chlorinated solvents). Benzene, saturated hydrocarbons, or tetrachloroethylene require days to be significantly oxidized by molecular ozone.

In bromide-containing waters (seawater, brackish waters, pool waters), the bromide is the primary reactant with ozone. It is oxidized to hypobromite or even bromate, and can destroy part of the ozone in a catalytic chain process. The hypobromite can produce bromoform.

Some of the aqueous ozone always decomposes. This decomposition is often initiated by increased pH (OH⁻-ions) and accelerated by many types of organic compounds which may act as promoters of a radical type chain reaction. This chain reaction is, however, somewhat inhibited by carbonate. Therefore, the half-life of ozone varies according to the type of water treated. For example, in typical drinking water it can be anything from 1 to 20 minutes.

Ozone decomposition leads to the formation of the OH· radical. This is so reactive that, in drinking water, most of it is consumed by humics and other dissolved organic materials. The OH· radical can even be reduced by carbonate and bicarbonate ions. But a fraction of this oxidant is still available to oxidize a specific micropollutant, even when this is present in relatively low concentrations. Based on known kinetic constants, all these effects are easily predictable.

UV-irradiation of aqueous ozone produces primarily hydrogen peroxide which, under certain circumstances, may initiate decomposition of additional ozone to OH· radicals.

The chemical effects of ozonation processes must also be considered with respect to combined processes, where ozonation is, e.g. followed by treatment with chlorine or microbiological treatment steps, or even by membrane filtration in the case of industrial waste water treatment.

1. Introduction

This chapter summarizes some selected characteristics of ozone chemistry with respect to water treatment processes.

Process engineers who want to apply ozone most effectively must take into account the chemical characteristics of ozone. First, one of the many potential applications of ozone may be neglected due to lack of knowledge of the specifics of ozone chemistry and their potential to solve given problems. Further, in ozone application, much time, money, and goodwill have been lost in the past, and will continue to be lost in the future, by ignoring some fundamental rules of ozone chemistry. For these reasons, the main chemical characteristics of an ozonation process should always be reviewed before planning and performing experiments to optimize an application. In the following section it also becomes clear that the chemical-engineering part of an ozonation process can still be greatly improved when the kinetics and mechanisms of ozonation reactions are taken into consideration.

As depicted in Fig. 1, standard ozonation treatment must be effective at ozone concentrations of only a few mg/l, within relatively short reaction times of less than 1000 s, and at low ambient

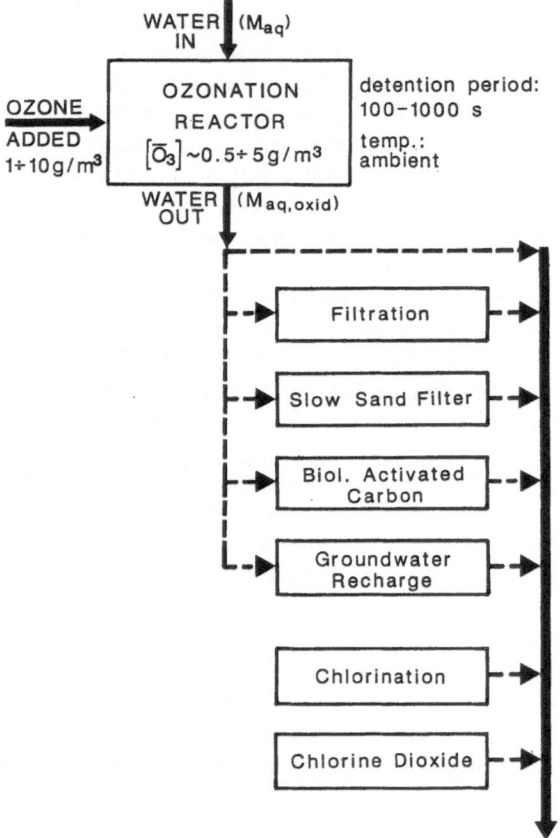

Fig.1:   The conventional ozonation processes for drinking water and waste water treatment are performed by dosing a few g ozone per m³ of water (lower values suffice in the case of good drinking water). The detention time of the water in the reactor is typically in the range of 100 to 1000 s.
The ozonation process is generally followed by further treatment steps. Most surface water derived drinking water requires at least post-treatment which allows for microbiological degradation of the ozonolytic products, e.g. those formed from dissolved humic materials.

temperatures of water. In addition, ozone is actually a very weak oxidant, so that it is, e.g. applied by organic chemists only when they want to oxidize selectively very special chemical configurations within larger molecules (ozonolysis). It must, therefore, be kept in mind that, for most applications of ozone, the strength or efficacy of the ozonation process is a direct consequence of the weak oxidizing power and, therefore the great selectivity, of ozone. It is because of this selectivity that most water treatment processes can be performed with relatively low ozone dosages, which are sufficient to improve the hygienic as well as physio-chemical qualities of the treated water. A more reactive oxidant would be consumed in large amounts in arbitrary (non-efficacious) oxidation reactions. In only a few applications may it be advantageous to decompose ozone to produce more reactive (and less selective) secondary oxidants, such as OH' radicals.

For processes requiring higher ozone doses, the cost of this oxidant, its low water solubility combined with its low mass transfer, and its tendency to decompose under certain conditions, restrict its use.

As indicated in Fig. 1, ozonation generally comprises a single step in a long sequence of treatment processes[1]. Therefore, the chemistry of ozonation must be optimized, not in isolation, but in the context of the entire treatment sequence.

As shown in Fig. 2, the chemical effects of ozone in water are a result of:

- its direct reactions with dissolved compounds (M),

- its decomposition into secondary oxidants, such as highly reactive radicals (OH', $HO_2^.$),

- the formation of additional secondary oxidants from ozone reacting with other solutes, e.g. the formation of HOBr (active bromine) when ozone oxidizes bromide ions,

- the subsequent reactions of these secondary oxidants with solutes (M).

All these reactions may occur simultaneously. In practice, however, one or the other reaction will predominate, depending on the reaction conditions and the chemical composition of the

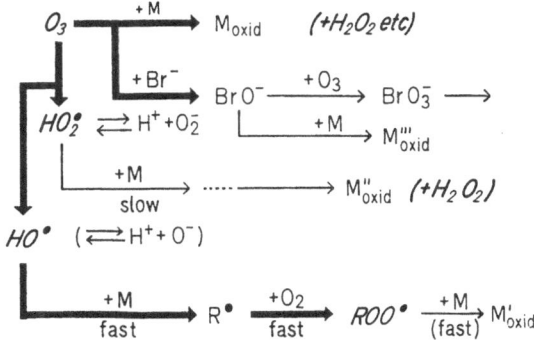

Fig. 2:   This scheme shows the different pathways of the reactions of ozone with solutes (M) (including Br'), and the formation of secondary oxidants, which also react with solutes M, but which produce different products from those formed on direct reaction with ozone.

water being treated. Accordingly, the reactions given in Fig. 2 can be separated into subunits which are discussed and interpreted individually below.

2. The Direct Reaction of Ozone with Dissolved Compounds

Thermodynamically, ozone is a very strong oxidant. However, most of its reactions are so slow that its chemical effects are controlled by kinetics, and not by thermodynamics.

Experience shows that all primary reactions of ozone with dissolved compounds can be expressed by the same simple rate law, where the rate is first order with respect to both the concentration of ozone $[O_3]$, and the concentration of the solute $[M]$. So, for the reaction of O_3 with M, we can write: $O_3 + M \xrightarrow{k_M} M_{oxid}$ and the rate of disappearance of M in the presence of a given concentration of ozone $[O_3]$ becomes:

$$-\frac{[M]}{dt} = k_M [M]^1 [O_3]^1 . \tag{1}$$

If we consider a plug-flow or a batch-type reactor, the rate-law for the elimination of M can therefore be written as:

$$-\ln \frac{[M]}{[M]_0} = k_M [O_3] \cdot t . \tag{2}$$

Thus the logarithm of the relative residual concentration of M declines linearly with the

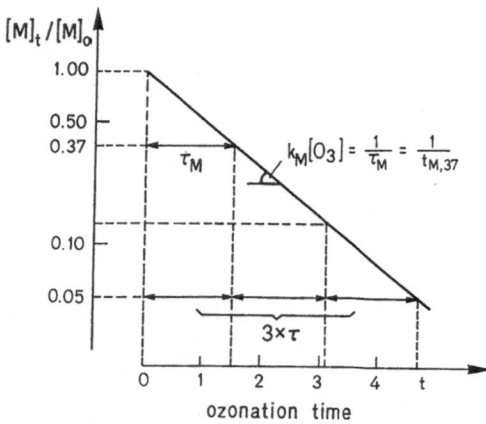

Fig. 3:   Relative residual concentration (ln scale) vs. time of a solute M which reacts with ozone with a rate constant k_M (compare equation (2)). Assumptions:
  • constant ozone concentration during the process time,
  • batch-type or ideal plug-flow reactor,
  • M is an individual chemical species, and not a sum parameter.

τ_M ($= t_{M,37}$)   is the time required to reduce the concentration of M by a factor e (i.e., to 37%), see also equation (3).

time t during which a given concentration of ozone is present (see Fig. 3). The slope increases with the concentration of ozone and the substrate-specific reaction-rate constant k_M. This kinetic law shows that the extent of the direct reaction is proportional to the product of concentration of ozone multiplied by the reaction time.

For practical applications, it should be emphasized that the efficiencies of these direct reactions of ozone with micropollutants are only defined when the relative decrease of the concentration of the micropollutant is based on the mean concentration of the ozone acting during the stated process time.

The rate constants for the direct reaction of ozone with organic solutes in water are published for about 110 representative organic substances and for about 60 inorganic species[2-4]. Recently, also a list comprising carbohydrates (sugars) has been published[5]. In addition, rate constants published for non-aqueous media can sometimes be converted to the water phase using the relevant conversion factors which depend on the specific class of compounds[2]. However, in water many relevant compounds dissociate into ions or become complexed by cations, thus forming new aqueous species with very different rate constants[3]. Because experimental methods for determining reaction-rate constants for aqueous organics have now been developed and well tested, constants for additional compounds of interest may be easily measured in any laboratory. It is now, therefore, most appropriate to measure additional rate constants within case studies of applications.

In Figs. 4 to 6, a few examples of organic micropollutants and inorganic compounds are

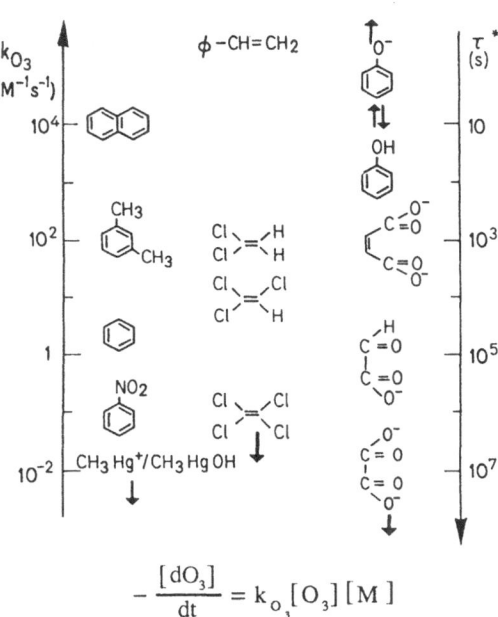

$$-\frac{[dO_3]}{dt} = k_{O_3}[O_3][M]$$

Fig. 4:   Examples of rate constants for direct reactions of ozone with organic solutes)
Right-hand scale: $t_{M,37}$ is the reaction time required to reduce the concentration of the solute by a factor e (i.e. to 37%) if the ozone is present at a concentration of 10^{-5}M (0.5 mg/l), batch type or plug-flow reactor[2,3].

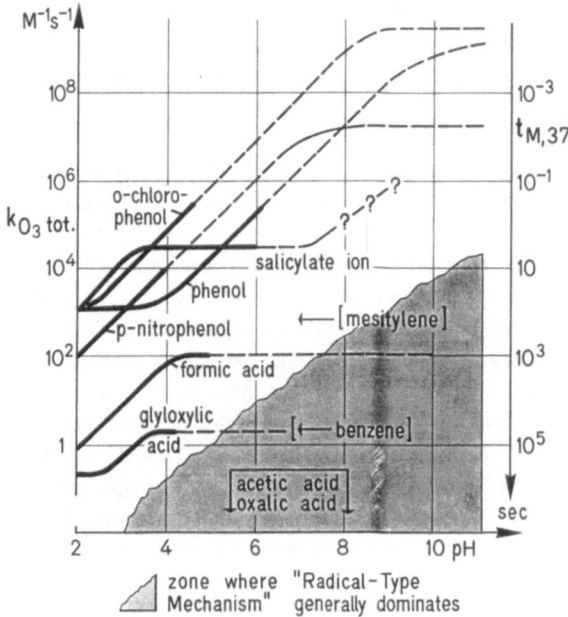

Fig. 5:   Rate constants for direct reactions of ozone with organic solutes vs. pH.
           Right-hand scale: $t_{M,37}$ is the reaction time required to reduce the
           concentration of the solute by a factor e (i.e. to 37%) if the ozone is present
           at a concentration of 10^{-5}M (0.5 mg/l), batch type or plug-flow reactor[2,3].

arranged on a scale based upon their second-order reaction-rate constants. On the right, a
corresponding scale indicates the time t_{37} within which the micropollutants are eliminated by
a factor e (i.e. to 37 %), assuming an O_3 concentration of 10^{-5} M (about 0.5 mg/l, typical for
drinking water treatment). For a concentration of ozone 10 times higher, the t_{37} values would
be only 1/10 as long. Thereby:

$$t_{37} = \tau = \frac{1}{k_M[O_3]} . \tag{3}$$

In these figures, for the conversion of k_{O3} to k_M values, the stoichiometric yield factor μ
($\Delta M/\Delta O_3$) was approximated by 1.0, although real values would range between
0.2 to 1.0 $\Delta M/\Delta O_3$. (Appropriate corrections would not show up using this compressed scale
which covers 10 orders of magnitudes.)

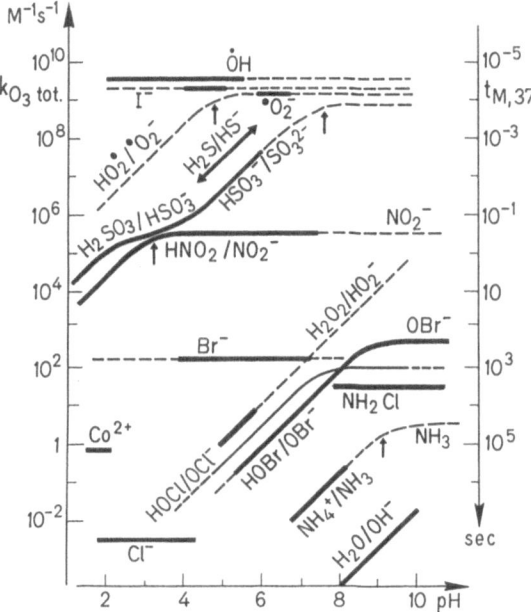

Fig. 6:   Examples of rate constants for direct reactions of ozone with inorganic solutes vs. pH (data selected from Ref. 4)
Right-hand scale: $t_{M,37}$ is the reaction time required to reduce the concentration of the solute by a factor e (i.e. to 37%) if the ozone is present at a concentration of $10^{-5}M$ (0.5 mg/l), batch type or plug-flow reactor[2,3].

From the few selected examples for organic compounds we learn that:

• Saturated alkyl-groups, such as present in alkanes (e.g. hexane) do not react.

• Olefinic compounds (e.g. oleic acids or styrene) react within seconds. However, when the H atoms in α position to the C-C double bond are replaced by chlorine atoms, even these unsaturated compounds become inert. Therefore, perchloroethylene or trichloroethylene, which are common contaminants in ground water, cannot be oxidized by this direct reaction within a reasonable ozonation time.

• Benzene reacts only within days. However, polyaromatic hydrocarbons (systems of condensed benzene rings which are considered to be carcinogenic) mostly react within seconds.

• Phenols react within seconds. The phenolate anion reacts 10^6 times faster than the non-dissociated phenols. Thus, even above pH 4, where at least 10^{-6} parts of the phenol are dissociated, the rate of reaction depends on the small amount of this phenolate anion, so that above pH 4 the apparent rate constant increases with the degree of dissociation, i.e. by a factor of 10 per pH unit. Such pH dependencies must be considered when the relative rates with which two substrates are eliminated during an ozonation process are compared at different pH values[3]. They also become important when mass transfer of ozone is accelerated by chemical reactions[6].

- The oxidation products produced by ozonolytic cleavage of aromatic ring systems are glyoxylate-, maleic-, oxalate-, acetate-, or formate ions (for a comprehensive recent review see, e.g. Ref. 7). Of these products, only the formate ion reacts slightly during further ozonation. All the others will accumulate as final products when no alternative pathways for oxidations are operative (see Ref. 8 and 9 and the entries in Figs. 4 and 5).

From these findings we may conclude that, for water treatment applications:

- Direct ozonation reactions are highly selective. Only those compounds containing functional groups that are easily attacked by the electrophilic ozone become oxidized.

- The direct reactions of ozone cannot be applied to oxidize solvent-derived pollutants such as alkanes, benzene, or chlorinated organic compounds. However, the application of ozone should be considered in degrading poly-aromatic hydrocarbons, phenolic compounds, free (non-protonated) amines or sulfides. These latter compounds are often coloured or malodorous. Because these exhibit such high reactivities towards ozone, many waterworks use ozonation to remove malodorous and coloured contaminants.

The products of ozonation of organic compounds are usually more polar chemicals (e.g. acids), which are more water soluble, less volatile and less lipophilic. They therefore tend to be both less odorous and less toxic. However, ozonolysis may also break down larger organic compounds containing olefinic groups into smaller and sometimes more volatile ketones and aldehydes (classical ozonolysis). In natural surface waters, such intermediates may produce a somewhat acidic (citronic) smell, which is only overcome through prolonged ozonation during which these compounds undergo further oxidation.

Of the inorganic species given in Fig. 6, sulfide (HS^-), sulfite (HSO_3^-), nitrite (NO_2^-) and iodide (I^-) anions react immediately. The oxidation of sulfide leads to sulfate in a fast reaction. Whenever sulfide is present, this reaction is so fast that it occurs immediately, before much ozone is consumed by the other pathways. The rate of this reaction is of importance when sulfide containing mineral waters are treated with ozone (this process is approved by the Swiss Office of Health, for example, in treating mineral waters). This reaction can also be used in the

Summary of Direct Ozone Reactions

1.	They are controlled by kinetics, and not by thermodynamics.
2.	All rates of reactions are first order in ozone and substrate concentration.
3.	Ozone is a highly selective oxidant.
4.	For many substrates, specific rate constants are listed in the literature. Laboratory methods are well established for measuring reaction rate constants over the range of 10^{-3} to 10^9 $M^{-1}s^{-1}$.
5.	Product formations change with ozone dose (ozone concentration multiplied by reaction time). The products mostly consist of a wide range of organic aldehydes, ketones, acids, and even polymers.
6.	The reaction kinetics and chemical rules for product formations are well-established but the final ozonolytic products formed from different substrates are only known approximately.

treatment of some waste water produced in scrubbing malodorous exhaust gases. Because ozone only attacks the free or non-protonated ammonia (NH_3), the apparent reaction-rate constant for ammonia increases with pH up to the pH region where all ammonium ions (NH_4^+) are dissociated. But even at such high pH (e.g., pH 9), thousands of seconds of ozonation time would still be required to oxidize a significant amount of this compound.

Nitrite can be oxidized quickly even in the presence of other solutes. This fact is relevant to processes where nitrite has been produced in biological activated filters, through anaerobic denitrification, and other water treatment processes leading to the formation of nitrite.

Aqueous chlorine also reacts at elevated pH where it is present as hypochlorite anion and not as hypochlorous acid. But even at pH > 7.5, the required time of reaction is still in the order of 1000 s.

At pH 8, chloramine is degraded somewhat slower than free chlorine. Compared with the ozonolysis of chlorine, this reaction is, however, not retarded when the pH is lowered. The ozonation of chloramine leads to the formation of chloride and nitrate. This ozonolytic reaction is applied in eliminating chloramines in swimming-pools in order to avoid the build-up of eye irritants (for a review see Ref. 11). Hence, it is known that extended ozonation treatments are required, which accords with the rate constants published for this reaction[12,13].

The oxidation time for bromide in the presence of 0.5 mg/l ozone is about 1000 seconds. The lifetime of small concentrations of ozone in water containing 2 mg/l bromide would be 500 s if the ozone were consumed only by this reaction. Seawater contains 65 mg/l bromide which consumes ozone within about 5 seconds. This interaction between ozone and bromide becomes the dominant reaction when brackish water or seawater are ozonated[14]. But the oxidation of bromide must also be considered in bromide-containing drinking water. (Some drinking water contains bromide concentrations in the 0.1 to 2 mg/l range, for example, locations below alkali mines such as those situated along the Rhine below Seltz or around Barcelona, some ground water in Northern Germany and Belgium, the main water supply in Israel (from the Lake of Galilee), and some mineral water.)

As shown in Fig. 7, ozone reacts with bromide to produce bromine (HOBr), which often appears as a residual oxidant but which can also be slowly further oxidized either to bromate or to a peroxide. The latter decomposes and reforms bromide. Bromide therefore decomposes either part of the ozone in a chain reaction, or converts it into bromate. Alternatively, the intermediate bromine can also react with ammonia or with organic substrates. In principle,

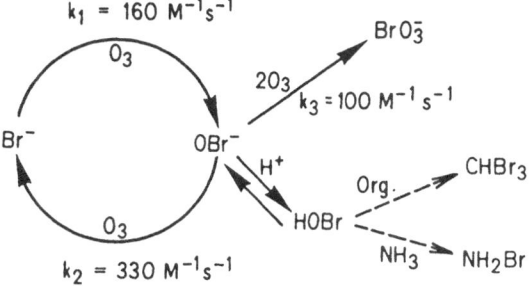

Fig. 7:   Reactions of ozone with bromide and hypobromite[12].

similar sequences of reactions are expected to those which are observed when chlorination of bromide-containing waters produces bromine. From experience using chlorine (for a recent review see Ref. 15) we can anticipate what types of secondary water contaminants are also to be expected for the ozonation process. But in the case of ozonation, the bromine becomes further oxidized and less bromine accumulates during ozonation than during chlorination.

In experiments some formation of brominated organic compounds, such as bromoform, has still been observed during ozonation of bromide-containing waters[12,13,16,17]. In actual water treatment applications it seems, however, that significantly less bromoform is produced through ozonation treatment than through chlorination of bromide-containing drinking water. Apparently, in such treatments, ozone also interacts with short-lived precursors of the halo-form formation process (see Ref. 18 and Ref. 19 p.175). If, however, bromide is added to swimming pools in order to increase the disinfecting power of ozone by producing bromine as a residual, much bromoform may accumulate where organic contaminants are present. An ozonation/bromide process must therefore be viewed with circumspection[20].

When phenols are present in bromide-containing water, the formation of bromophenolic compounds is to be expected. However, ozone degrades phenols (including bromophenols) much faster than bromine is produced from bromide as shown by the entries in Fig. 5 and 6. Therefore bromophenols do not persist in ozonation processes, whereas when these compounds are formed during chlorination they make the water taste like an unventilated pharmacy.

Warning: The kinetics of an elimination process can only be quantified if individual specified substances of known chemical speciations (degree of dissociation) are considered (exceptions are where the kinetic order of reaction is zero). Group parameters, such as colour or total light absorption values, total organic carbon content (TOC), and concentration of total organic halogen compounds (TOX) are lumped parameters which represent a mixture of individual substances with various reactivities. The kinetics of the transformations of such group parameters depend therefore on their actual composition. Because this composition changes as the reaction proceeds, a non-defined shift in the apparent rate constants is observed during the process time. In other words, an "aging of the kinetics" occurs. Thus, no general kinetic laws and constants based on group parameters can be formulated. Kinetic systems or oxidation processes must therefore be calibrated by observing specified reference substances. In this respect, the art of experimental practice is not different from that employed in disinfection studies, where calibrations must also be based on well-defined organisms.

Summary of the Effects of Ozone in Bromide Containing Water

1.	Bromide is slowly oxidized to "aqueous bromine" (HOBr/OBr⁻), which may be:
	• further oxidized to bromate,
	• further oxidized to an intermediate which reforms bromide,
	• added to organic solutes, part of which can produce bromoform,
	• added to ammonia.
2.	The oxidation of bromide to bromine occurs relatively slowly. Hence all phenolic compounds present in the water are ozonated before bromophenols are produced. This is different in chlorination processes, where bromophenols appear as noxious products.

3. Oxidation by OH· Radicals

Aqueous ozone decomposes. Half of the ozone introduced into raw water from Lake Zurich (pH ~ 7.8, [HSO$_3$] ~ 1.2 mM),for example, decomposes within the ozonation time of 10 minutes, even after the spontaneous ozone demand of the water of less than 0.1 mg/l ozone has been met. In ground water containing low concentrations of dissolved organic material (DOC), the lifetime can be about twice this value, but in many other types of water, decomposition is even faster[21]. About half of the decomposed ozone is converted to the OH· radical[22], which is the most reactive aqueous oxidant. OH· can act on organic micropollutants by either H-abstraction, or OH· addition to a double bond, or an electron-transfer reaction (see Fig. 8). The resulting radicals easily add to the oxygen molecule (a "biradical"), which, in aerated water, is present in relatively high concentrations. The resulting peroxy radicals disproportionate or combine with each other, forming many types of mostly labile intermediates which react further to produce peroxides, aldehydes, acids, hydrogen peroxide etc. The product formations depend on so many different parameters that our understanding is, necessarily, restricted to a few selected examples on which detailed chemical analysis have been performed. The results of such exemplifications can only be generalized if the extent of OH· radical processes occurring can be calibrated or predicted.

A kinetic analysis of OH· reactions can be summarized as follows: OH· radicals are so reactive that they are rapidly consumed (within microseconds) by the sum of bicarbonate and all the organic compounds present in the water. Only a few will survive long enough to react with a specified micropollutant. In such a situation, the kinetics with which a specified micropollutant M is eliminated in a batch-type reactor can be described by:

$$\ln \frac{[M]}{[M]_0} = -\eta(\Delta O_3) \frac{k_M'}{\sum_i k_i'[S_i]} \cdot \tag{4}$$

This function is depicted in Fig. 9.

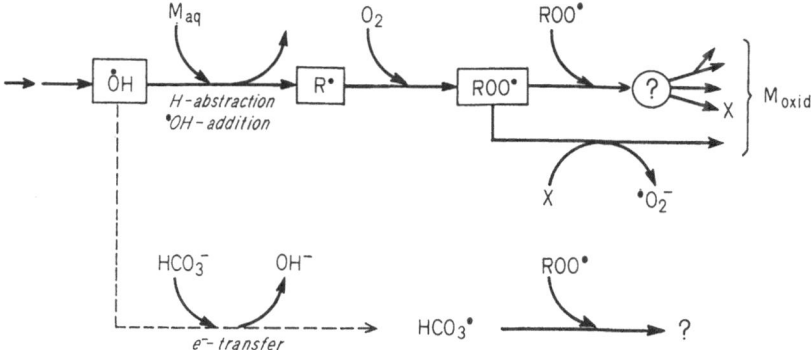

Fig. 8:   Reactions of OH· radicals produced from decomposed ozone (ΔO_3) with an organic solute M. A secondary radical (R) is produced which reacts with oxygen (O$_2$) to form a peroxy radical (ROO·). Such peroxy radicals undergo a series of further reactions until a stable end product appears.

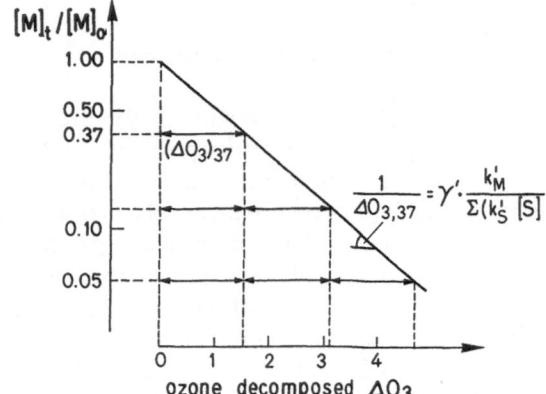

Fig. 9:   Relative residual concentration of an ozone-resistant reference
compound M plotted vs. the amount of ozone which is decomposed to
form OH· radicals (ΔO_3). Assumptions:
- the rate at which the sum of all solutes scavenge OH· radicals does not
  significantly change during ozonation,
- batch-type or ideal plug-flow reactor,
- M is an individual chemical species and not a sum parameter.

Thus, the logarithm of the relative residual concentration of M declines linearly with $\eta(\Delta O_3)$, the amount of ozone decomposed to OH· radicals. The elimination of M increases with the rate constant with which OH· reacts with M, (k'_M), but it decreases with the rate with which the sum of all other substrates, S_i, scavenge OH·, i.e. with $\Sigma(k'_i[S_i])$.

Fig. 10 shows the depletion of ozone resistant micropollutants in water from Lake Zürich which react only with these secondarily produced OH· radicals. Here toluene was spiked into an additional sample as a reference solute. Its relative rate of depletion roughly corresponds with that of micropollutants present in 10^4 times lower concentrations. In Fig. 11, the relative rate of depletion of an ozone resistant micropollutant, which was added to a sample of water from a more eutrophic lakewater is shown. Based on equation (4) and considerable experience, (see above), the effect of OH· radicals on further micropollutants can easily be calibrated by analyzing the depletion of any probe substance which is not oxidized by ozone in a direct reaction. Conversion factors can be determined from the many lists of rate constants for OH· radical reactions published in the literature (for a review see Ref. 23). None of the experiments performed so far on more than 50 different types of water, using a variety of OH· probe molecules, have contradicted these results (for an early overview see Ref. 21). OH· radical reactions are not disturbed by particles and the same rates for the elimination of OH· probe molecules have been found prior to and after filtration of lakewater (see e.g. Fig. 11).

Some examples of rate constants of OH· radicals are presented in Fig. 12. They generally range from 10^9 to 10^{10} M^{-1}s^{-1}. On the right-hand scale of this figure, the kinetic values have been converted to values which indicate the amount of ozone which must decompose to achieve an elimination by a factor of e (to 37 %):

$$(\Delta O_3)_{37} = \frac{\Sigma(k_i[S_i])}{\eta \cdot k_M} \tag{5}$$

(In our earlier publications we called $(\Delta O)_{37}$ the "Ω-value").

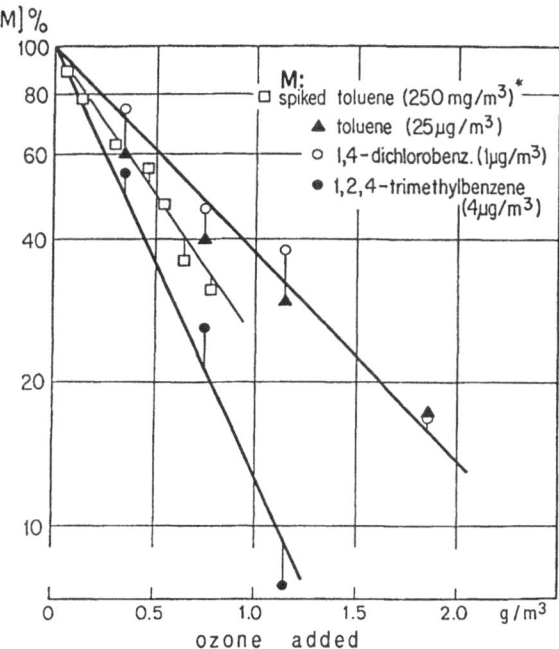

Fig. 10: Elimination of ozone-resistant organic micropollutants in water from Lake Zürich, following the decomposition of ozone to the more reactive OH˙ radical (only trimethylbenzene was also directly oxidized by ozone itself). The elimination of the micropollutants is compared with that of spiked toluene which was used as a reference at a 10,000 times higher concentration[10].

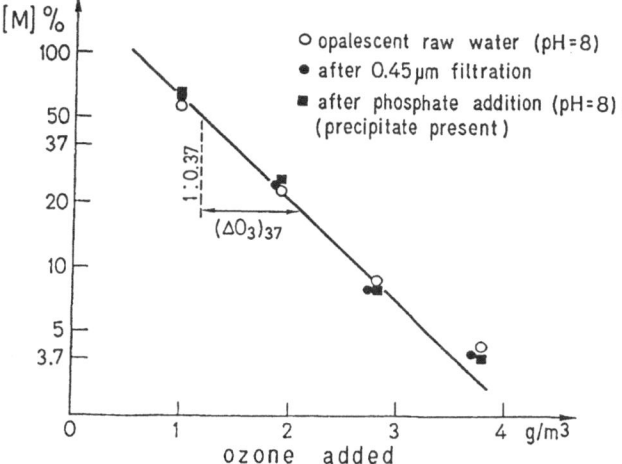

Fig. 11: Elimination of an ozone-resistant organic reference compound (M), benzene, which had been spiked to samples of water from the eutrophic Lake Greifensee (DOC ~ 4 mg/l; pH ~ 8). The reaction was shown to proceed via OH˙ radicals. The presence of particles did not inhibit the reaction[2,3].

This right-hand scale is based on a scavenging rate for OH· radicals which is typical for an eutrophic type of lakewater. It increases proportionally to the DOC, which here controls the lifetime of OH· radicals.

It is interesting to note that, in the surface water used for calibrating the right-hand scale in Fig. 12, about 1 to 10 mg/*l* of ozone had to be decomposed to eliminate most types of ozone resistant organic micropollutants by OH· radical reactions by a factor e (to 37%). For lakewater from Lake Zürich, which contains about three times fewer OH· radical scavengers, about three times less decomposed ozone would be required to achieve a comparable result.

Because the direct reaction of ozone and the reactions succeeding the formation of OH· radicals produce different products, the two pathways have to be distinguished and their relative importance considered whenever product formation is being discussed.

Summary of OH· Radical Reactions

1.	In most water OH· radicals are the main secondary oxidants produced from decomposed ozone. They are the most reactive oxidants known to occur in water.
2.	Most of the OH· radicals are produced in chain reactions where OH⁻ (high pH) and HO_2^- (dissociated hydrogen peroxide) act as initiators. Some organic solutes act as promoters, but bicarbonate and some types of organic compounds act as inhibitors of this chain reaction. In general, about 1/2 as much OH· radical is produced as O_3 decomposed.
3.	In natural water, most OH· reacts with bicarbonate or non-specified organic solutes, which both act as "scavengers". But a small fraction of OH· survives until it reacts with specified micropollutants. The reaction of OH· with a specified micropollutant becomes faster the higher its concentration and the cleaner the water, that means, the lower the concentration of the scavengers.
4.	Product formations from OH· radical reactions with micropollutant have been identified in only a few cases. In general, the products resulting from these radical reactions are too diverse and too complex to identify and quantify the individual species formed.

4. Rate of Decomposition of Aqueous Ozone

From experience, we know that the lifetime of ozone in water is shorter the higher the pH. But in most water decomposition also occurs in radical-type chain reactions, which can be somewhat inhibited by bicarbonate. Ozone is therefore more stable in water containing more bicarbonate (see Fig. 13) and the lifetime of ozone in good ground water can be longer than in highly distilled water of comparable pH. Even some special organic compounds can act as strong inhibitors for the ozone chain-decomposing reaction. But other types of organic impurities, e.g. carbohydrates or even methanol also promote the radical-type chain reaction, and hence accelerate the decomposition of ozone.

These empirical observations can be rationalized when considering the complex scheme of reactions given in Fig. 14 which was based on detailed experimental research and also tested by further experiments. This scheme shows that the rate of decomposition of ozone and the

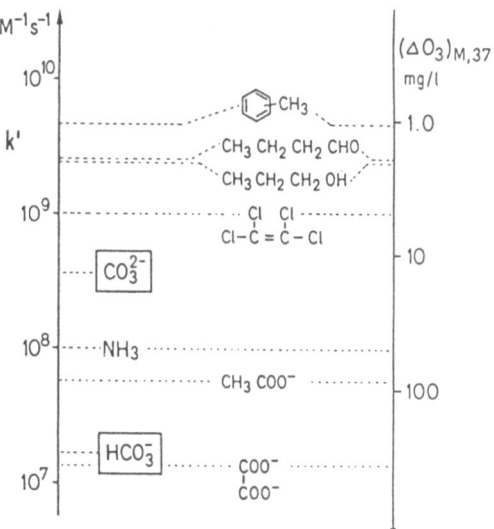

Fig. 12: Rate constants for reactions of OH' radicals with different solutes. $(\Delta O_3)_{37}$ is the required amount of decomposed ozone, which results in the elimination of the quoted substrate to 37 % of the initial value (batch-type or plug-flow reactor). This scale is calibrated for eutrophic lakewater (Lac de Bret, DOC = 4 mg/l, [HCO$_3^-$] = 1.6 mM, pH = 8.3. The latter changes proportionally to the DOC of water).

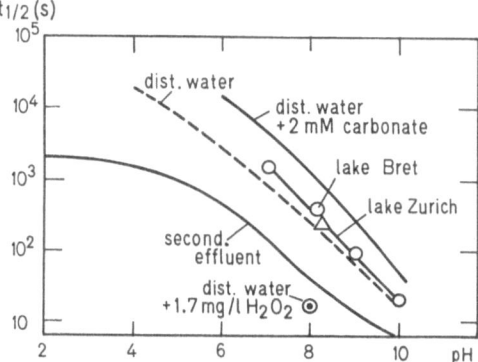

Fig. 13: Second half-life of ozone ($t_{1/2}$) in different types of water.

70% lakewater Zürich: DOC = 12 mg/l, 70 % lakewater Lac de Bret (DOC = 3.2 mg/l), 70% secondary effluent from municipal waste water, (DOC = 7 mg/l) borate buffer (the second half-life is defined as the time within which the concentration of ozone (about 3 mg/l) declines from 50% to 25% of its initial value (from Ref. 27 with some additions).

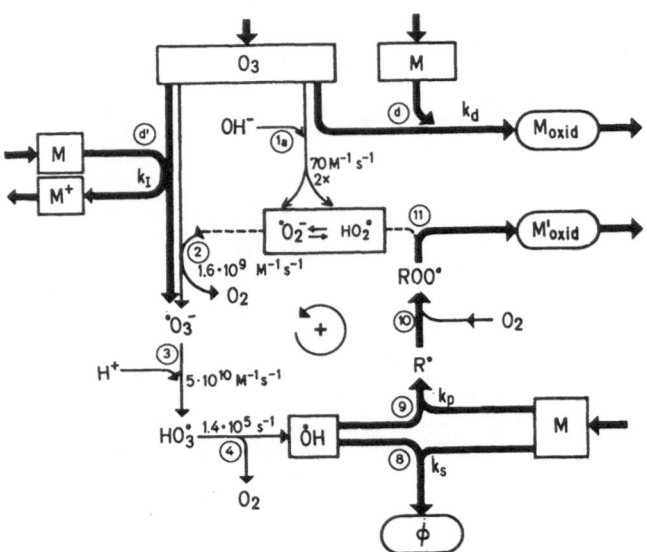

Fig. 14:   Reactions of aqueous ozone in the presence of solutes M which react
             directly with ozone (d), or which react with OH· radicals (reactions
             8 or 9). In reaction 8, M scavenges OH· without producing a chain carrier.
             In reaction sequence 9 to 11, however, M acts as a promoter for a chain
             reaction by converting OH· radicals to HO_2· which selectively reacts with
             additional ozone. In this case, ozone is consumed in a radical-type chain
             reaction in which OH· and HO_2· act as chain carriers and M as a promoter.
             In cases where the concentration of ozone relative to M is high, ozone
             itself can also act as a promoter by converting OH· into HO_2·[26].

formation of OH· radicals depends on many reaction sequences. The depicted chain reaction
can be initiated by OH⁻ (reaction 1a) which produces HO_2·/O_2⁻. It therefore often proceeds
more rapidly with higher pH. In addition, many other solutes, even hydrogen peroxide, which
are present in real water can act as initiators (reaction d').

The ozonide anions, ·O_3⁻ formed by electron-transfer from O_2⁻ to O_3 (reaction 2), decompose
upon protonation (reaction 3) to OH· radicals[24], which are consumed by organic compounds,
M. Some intermediates formed release O_2⁻ (typical reactions are when M = alcohols, sugar, or
formic acid), which transfers an electron to additional ozone in a selective reaction. This means
that solutes which transform non-selective OH· radicals into O_2⁻ act as promoters of the chain
reaction (reaction 9). In very clean water and in the presence of relatively high concentrations
of ozone, ozone itself can also act as a promoter by converting OH· into O_2⁻ (OH· + $O_3 \longrightarrow HO_4$,
which decomposes into O_2⁻)[25]. In contrast, other solutes can scavenge OH· producing
intermediates which cannot react with additional ozone (reaction 9). These solutes (e.g.,
bicarbonate) inhibit the chain reaction and thus stabilize the ozone. The situation is
complicated because the effects of different solutes do not contribute additively to the rate of
the chain reaction; rather, it is the relative amount of promoter to inhibitor which controls the
rate of decomposition of ozone to OH· by this chain reaction[26].

Because of these complexities, we recommend that the rate of decomposition of ozone for
a given water type be determined experimentally, and that the efficiency of the OH· radical
reactions be calibrated by observing an ozone-resistant OH· radical probe solute. For the

planning of good experiments and for the interpretation of the results, the kinetic characteristics of the chain reaction and of the OH· radical reactions must be taken into account[27]. The subsequent generalization of the results with respect to the oxidation of micropollutants is only possible if the tested water has been clearly characterized and the amount of decomposed ozone determined.

5. Combination of Ozone/UV and Ozone/Hydrogen Peroxide

UV light (255 nm from low pressure mercury lamps) is often applied in decomposing aqueous ozone. In water, this decomposition leads primarily to the formation of hydrogen peroxide[28], which is a rather inert oxidant. This aqueous phase reaction is very different from that which occurs in the atmosphere. Table 1 shows on how much hydrogen peroxide is accumulated during the UV decomposition of ozone in different types of water as determined experimantally. This hydrogen peroxide formed may decompose additional ozone into highly reactive OH· radicals. The scheme of reactions is presented in Fig. 15. The interaction of hydrogen peroxide formed with residual ozone occurs faster the higher the pH; so this reaction is only relevant if the pH is typically above 6 or when the UV intensity is very low. At lower pH values or at higher UV intensities, the ozone decomposes into hydrogen peroxide before this product has a chance to react with any residual ozone to produce OH· radicals.

A combined ozone/UV process therefore results in the formation of residual hydrogen peroxide, except when the process is performed at increased pH. This rather persistent product is detrimental for subsequent water treatment with chlorine or ozone because hydrogen peroxide can reduce both of these oxidants. It could also be detrimental for pharmaceutical products if the ozone/UV process is applied in the production of ultrapure water in the pharmaceutical industry.

Table 1: Typical yields for H_2O_2 formation from UV-photolyzed ozone in tri-distilled water
a) ++: UV intensity resulting in a half-life for ozone of 7s ($[O_3]_0 < 10$ μm) (Water exposed 8 cm from UV-low pressure lamps (0.14 W/cm)).
b) +: Lower UV intensity achieved by shielding with "Parafilm", resulting in a 10 fold longer lifetime for ozone.
(Extracted from Hoigné and Bader, Ref. 22.)

UV-irradiation	added $[O_3]_0$ (mg/l)	$\dfrac{\Delta H_2O_2(\text{Mole})}{\Delta O_3(\text{Mole})}$
++[a)]	0.015 - 0.05	0.8 - 1.0
	1.5	0.8
	15.0	0.5
+[b)]	0.15	0.7
	1.5	0.5
	15.0	0.1

Because of these chain reactions, the same amount of OH· radicals seems to be produced when either hydrogen peroxide or hydroxide anions act as initiators in decomposing ozone[22]. The effect of the OH· radicals produced in these combined processes is therefore comparable to that described above where OH· radical production was initiated by high pH. Only at relatively high concentrations of hydrogen peroxide does this initiator itself significantly scavenge OH· and thereby inhibit the OH· effect. But, in the UV/ozone process, a fraction of the ozone is always used for the formation of hydrogen peroxide and the formation of OH· radicals is hence reduced by this fraction.

Examples of the effect of such combined ozonation/UV processes are discussed in Ref. 29 and 30, which also give an up-to-date literature review. Examples for combined ozone/ hydrogen peroxide are to be found in Ref. 31 and 32.

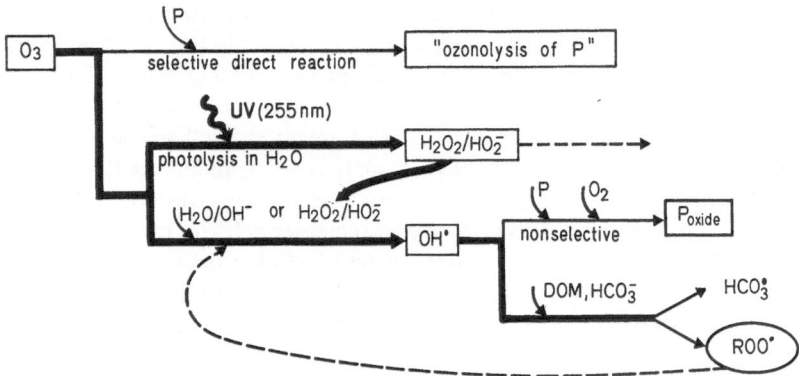

Fig. 15:   Pathways for transformation of ozone by reactions with pollutants (P) or OH⁻ or H_2O_2, or by UV photolysis [22].

Table 2:   Formation of trichloronitromethane (TCNM) (chloropicrin) and chloro-form (trichloromethane ($CHCl_3$)) in Greifensee water (TOC ~ 4 mg/*l*; pH = 7.8) when pre-ozonation is followed by chlorination.
   *) Doses of added ozone or chlorine. Chlorine was only added after all the ozone had reacted. Residual chlorine was eliminated one day after its dosage. *(Extraction from Hoigné and Bader , Ref.34, Table 4)*

treatment		product formation	
*) O_3 (mg/*l*)	*) Cl_2 (mg/*l*)	TCNM (µg/*l*)	$CHCl_3$ (µg/*l*)
-	-	0.02	0.2
-	2	2	45
1.0	2	4	30
2.0	2	6	28

6. Combination of Ozone with Other Disinfectants

Ozone interacts with chlorine dioxide and other disinfectants when these are applied simultaneously in combination processes. Richard and Brener[33] extensively discussed the interaction between ozone and chlorine. A kinetic analysis of the system is reported by Haag and Hoigné[13]. A simple overview, including other interactions as well, has been published recently by Hoigné[19].

7. Role of the Chemistry of an Ozonation Process for Forthcoming Treatment Steps

In most cases, normal practice is for ozonation to be succeeded by at least one further operation. The effects of the following subsequent treatment steps have consequently been investigated most intensively (see also Fig. 1):

- rapid sand filtration, where pre-ozonation process may act as a flocculant,
- flocculation,
- biological activated carbon filtration, or slow sand filtration, or ground water recharge, for which pre-ozonation forms further biodegradable products,
- chlorination,
- chlorine dioxide addition,
- chloramine addition.

Most of these chains of processes have been discussed at recent ozone conferences and comprehensive surveys can be found in corresponding proceedings (also see Ref. 29). In the case of post-chlorination, we would like to emphasize that, on the one hand a pre-ozonation can lead to some decreased haloform formation, but the extent of this effect depends largely on process- and water parameters; on the other hand, in surface water, such a process combination also leads to the formation of some chloropicrin (nitro-chloroform). An example of results is given in Table 2. This chloropicrin formation does not seem to be of hygienic concern. It could, however, be an indicator of the formation of further nitro-chloro compounds, which might complicate the hygienic situation. If this last example is mentioned, it should be emphasized that there are still open questions. However, wholesomeness tests on ozonated drinking waters have not shown any detrimental results so far.

For drinking water distribution, the replacement of post-chlorination by an addition of chloramine seems to be recommended whenever the viricidic action of the pre-ozonation can be guaranteed. Wherever ozonation can be applied, it seems to be the preferable oxidation treatment, especially when it can be succeeded by a microbiological treatment step, which eliminates some of the biodegradable ozonation products before these give rise to fouling processes in the distribution system.

8. Analysis of Aqueous Ozone

In order to predict the rate of an ozonation reaction, either the absolute concentration of ozone which acts during the process time must be known, or the process must be calibrated by following the change of concentration over time for a probe molecule of known reaction-rate constant. When ozone is applied in disinfection, a well defined microorganism can also be used for calibration.

The concentration of residual ozone is most easily determined by measuring the UV absorption at 258 nm, where aqueous ozone has an absorption coefficient of 2950 ± 50 $M^{-1}cm^{-1}$. This value has been reinvestigated using a better characterized primary chemical standard (NO_2^- system). But many aqueous solutions themselves absorb significantly in the UV range. In these cases, ozone cannot be analysed by direct UV-absorption measurement, but the ozone concentration can be easily determined using the indigo-trisulfonate method. This method is based on the ability of ozone to discolour indigo reagent, which results in a change in absorbance at 600 nm of 20'000 cm^{-1} per mole/l of added ozone (this method is now accepted as the standard method in Switzerland and under consideration as a standard method in Germany and the U.S.)[35].

References

1 W.J. Masschelein, ed., *L'Ozonation des Eaux: Manuel Pratique* (Technique et Documentation, Paris) 1982. Engl.: *Ozonation Manual for Water and Wastewater Treatment* (John Wiley, New York) 1982.

2 Hoigné, J. and Bader, H., Rate constants of reactions of ozone with organic and inorganic compounds in water: I. Non-dissociating organic compounds. Water Res. **17** (1983) 173-183.

3 Hoigné, J. and Bader, H., Rate constants of reactions of ozone with organic and inorganic compounds in water: II Dissociating organic compounds. Water Res. **17** (1983) 185-194.

4 Hoigné, J., Bader, H., Haag, W.R. and Staehelin, J., Rate constants of reactions of ozone with organic and inorganic compounds in water: III Inorganic compounds and radicals. Water Res. **19** (1985) 185-194.

5 Perez, R.R., Gómez, M.M. and Ramos, L.R., "Ozone reactions with carbohydrates in aqueous medium" in *Proc. 8th Ozone World Congress,* Sept. 1987, Zürich. International Ozone Association (IOA). (Unionsverlag, Zürich) 1987, Vol. 2, pp. E106-127.

6 Masschelein, W.J. and Goossens, R., Nitrophenols as model compounds in the design of ozone contacting and reacting systems. Ozone Sci. Eng. **6** (1982) 143-162.

7 Ganducheau, Ch., Gilbert, E. and Eberle, S.H., Are the results of ozonation of model compounds at high concentrations transferable to the conditions of drinkingwater treatment with ozone? Ozone Sci. Eng. **8** (1986) 199-216.

8 Caprio, V., Insola, A. and Volpicelli, G., Ozonation of aqueous solutions of nitrobenzene. Ozone Sci. Eng. **6** (1984) 115-121.

9 Gaul, M.D., Junk, A.G. and Svec, H.J., Aqueous ozonolysis products of methyl- and dimethylnaphtalenes. Environm. Sci. Technol. **21** (1987) 777-784.

10 Zürcher, F., Bader, H. and Hoigné, J., "Verhalten organischer Spurenstoffe bei der Ozonung von Trinkwasser" in *Concerted Action Analysis of Organic Micropollutants in Water* (Cost Project 64 bis), Vol. 2, (Commission of the European Communities, Brussels) 1982, pp. 198-213.

11 Eichelsdörfer, D., "Application of ozone for treatment of swimming pool water in the Federal Republic of Germany" in *Proc. 8th Ozone World Congress*, Sept. 1987, Zürich. International Ozone Association (IOA). (Unionsverlag, Zürich) 1987, Vol. 2, pp. G 38-51.

12 Haag, W.R. and Hoigné, J., Ozonation of bromide-containing waters: Kinetics of formation of hypobromous acid and bromate. Envrionm. Sci. Technol. **17** (1983) 261-267.

13 Haag, W.R. and Hoigné, J., Ozonation of water containing chlorine and chloramines. Water Res. **17** (1983) 1397-1402.

14 Crecelius, E.A., Measurements of oxidants in ozonized seawater and some biological reactions. J. Fisheries Res. Board (Canada) **36** (1979) 1006-1008.

15 Rebhun, M., Heller-Grossmann, L., Manka, J., Kimel, D. and Limoni, B., "THM formation and distribution in a bromide rich and ammonia containing water" in *Water Chlorination 6, 6th Water Chlorination Conf.*, Oak Ridge, May 1987, R.L. Jolley, ed. (in press).

16 Maier, D. and Mäckle, H., Wirkung von Chlor auf natürliche und ozonte organische Wasserinhaltsstoffe. Vom Wasser **47** (1976) 379-397.

17 Cooper, W.J., Amy, G.L., Moore, C.A. and Zika, R.G., Bromoform formation in ozonated groundwater containing bromide and humic substances. Ozone Sci. Eng. **8** (1986) 63-76.

18 Kruithof, J.C., Noordsij, A., Puijker, L.M. and van der Gaag, M.A., "The influence of water treatment processes on the formation of organic halogens and mutagenic activity by post chlorination" in *Water Chlorination, Environmental Impact and Health Effects,* Vol. 5, R. L. Jolley, ed. (Lewis Publ. Inc., Chelsea) 1985, pp. 1137-1163.

19 Hoigné, J., Verhalten anorganischer Ionen and Desinfektionsmittel bei der Ozonung von Wasser (Übersicht). Gas-Wasser-Abwasser **65** (1985) 773-778.

20 P. Eichelsdörfer, Private communication, 1987.

21 Hoigné, J. and Bader, H., Ozonation of water: Oxidation competition values of 'OH radical reactions of different types of waters used in Switzerland. Ozone Sci. Eng. **1** (1979) 357-372.

22 Hoigné, J. and Bader, H., "Combination of ozone/UV and ozone/hydrogen peroxide: Formation of secondary oxidants" in *Proc. 8th World Conf. on Ozone*, Sept.19 87, Zürich. International Ozone Association (IOA). (Unionsverlag, Zürich) 1987, Vol. 2, pp. K 83-97.

23 Farhataziz, and Ross, A.B., "Selected specific rates of reactions of transients from water in aqueous solution; III Hydroxyl radical and perhydroxyl radical", NSRDS-NBS 59 (National Bureau of Standards, Washington, D.C., US Government Printing Office) 1977.

24 Bühler, R., Staehelin, J. and Hoigné, J., Ozone decomposition in water studied by pulse radiolysis. I. HO_2/O_3^- as intermediates. J. Phys. Chem. **88** (1984) 2560-64.

25 Staehelin, J., Bühler, R.E. and Hoigné, J., Ozone decomposition in water studied by pulse radiolysis. 2. OH and HO_4 as chain intermediates. J. Phys. Chem. **88** (1984) 5999-6004.

26 Staehelin, J. and Hoigné, J., Decomposition of ozone in water in the presence of organic solutes acting as promoters and inhibitors of radical chain reactions. Environm. Sci. Technol. **19** (1985) 1206-1213.

27 Hoigné, J., "Mechanisms, rates and selectivities of oxidations of organic compounds initiated by ozonation of water" in *Handbook of Ozone Technology and Application*, Vol. 1, R.G. Rice and A. Netzer, ed. (Ann Arbor Science, Ann Arbor, Mich.) 1982, pp. 341-379.

28 Taube, H., Photochemical reactions of ozone in solution. Trans. Farad. Soc. **53** (1956) 656-665.

29 Glaze, W.H., Drinking-water treatment with ozone. Environm. Sci. Technol. **21** (1987) 224-230.

30 Peyton, G.R. and Glaze, W.H., "Mechanism of photolytic ozonation" in *Photochemistry of Environmental Aquatic Systems*, Rod G. Zika and W.J. Cooper, eds., ACS Symp. Ser. 327 (ACS, Washington DC) 1987, pp. 76-78.

31 Brunet, R., Bourbigot, M.M. and Doré, M., Oxidation of organic compounds through the combination ozone-hydrogen peroxide. Ozone Sci. Eng. **6** (1984) 163-183.

32 Duguet, J.P., Broadart, E., Duissert, B. and Mallevialle, J., Improvement in the effectiveness of ozonation of drinking water through the use of hydrogen peroxide. Ozone Sci. Eng. **7** (1985) 241-257.

33 Richard, Y. and Brener, Y., "Interference between ozone and chlorine" in *Handbook of Ozone Technology and Application*, Vol. 1, R.G. Rice, and A. Netzer, ed. (Ann Arbor Science, Ann Arbor, Mich.) 1982, pp. 277-284.

34 Hoigné, J. and Bader, H., The formation of trichloromethane (chloropicrin) and chloroform in a combined ozonation/chlorination treatment of drinking water. Water Res. **22** (1988) 313-319.

35 Bader, H. and Hoigné, J., Determination of ozone in water by the indigo method; a submitted standard method. Ozone Sci. Eng. **4** (1982) 169-176.

Discussion

Chairman: M. Mirbach
Brown Boveri, Baden, Switzerland

C. von Sonntag •   Do you have any idea why you see an increase in the rate constant (Fig. 5) at about pH 6 for the reaction of phenol with ozone? The pKa of phenol, of course, is about 10.

J. Hoigné (referring to Fig. 5)   •   The reaction rate constant for phenol increases from 10^4 l/(mole sec) at pH ≈ 4 to 10^9 l/(mole sec) which is nearly diffusion controlled, at pH ≈ 10 corresponding to the pKa value of phenol. The phenolate anion hence reacts a million times faster than the non-dissociated phenol because of the negative charge which makes the ring more accessible to the reaction of the electrophilic ozone. Immediately beyond pH 4, 10^{-6} parts of the phenol are dissociated and the reaction rate becomes controlled by this one millionth. Increasing the pH by one unit leads to an increase of the degree of dissociation α by a factor of 10. The resulting curve has a unity slope on a log/log plot.
Whilst I have this diagram (Fig. 5), I would like to emphasize, for the benefit of the membrane specialists in the audience, that ozone treatment can effectively be combined with membrane technologies. For instance, the Swissair workshops at Zürich Airport have a process for membrane purification of their waste water which contains phenol among other things. Since the membranes deteriorate in the presence of phenol, a preozonation stage was introduced to remove the phenol selectively.

U. Kogelschatz •   You mentioned chloropicrin in your paper. Do you think this is of any importance in water treatment?

J. Hoigné   •   No, it is of no importance for the actual water treatment process as long as the consumer is not exposed to more than 1 µg/l. It has no effects on health. However, we take it as an indicator of the formation of other compounds we cannot analyze. If you combine it with an activated carbon filter, you can easily degrade chloropicrin.

J. Stankovic   •   Could you please tell us something more about the reaction order of ozone decomposition. Most authors assume a first order reaction. Are there other possibilities?

J. Hoigné   •   Thank you for this question. We have written quite a few publications on this subject and I think your question is difficult to answer! The situation is summarized in Fig. 14: the initiation reaction is just proportional to hydroxide anion concentration. The subsequent chain system depends on the relative amounts of scavengers and promoters present in trace quantities, which is the reason why all authors obtain different results. But, in principle, the rate of this chain increases linearly with the promoter concentration relative to scavanger concentration. In a highly scavanged solution, all these chain reactions are inhibited and ozone decomposition is first order in ozone and first order in OH⁻.

C. Kavanaugh   •   I seem to recall in some of your writings that the stoichiometric yield for hydroxyl radicals could be as high as two moles per mole of hydrogen peroxide which would suggest a possible improvement in the stoichiometric yield coefficient. Do I misunderstand that somehow?

J. Hoigné • Yes. Sorry, to seem contradictory - based on the reactions considered in Fig. 15 you would expect it to be one mole. Well, if you decompose ozone with hydrogen peroxide, you decompose one ozone molecule with HO_2^-. You thereby form one $OH^{\cdot}$ radical and one HO_2 radical which makes O_2^-. This decomposes another ozone. That is to say, you form from two ozone molecules two $OH^{\cdot}$ radicals. This corresponds to a 1:1 ratio. But in the real system, we only find 0.5. We assume now this is due to the fact that the radicals formed also consume another ozone molecule. But we obtain this very same yield independently of how we initiate the ozone decomposition. For details, I would like to refer to our last week's paper, Ref. 22.

C. Kavanaugh • I made a statement in my talk that the reaction of ozone with pollutants occurred so rapidly that the oxidation would take place in the liquid film. In the discussions later, Dr. Peyton from Illinois Water Survey challenged that and said that he didn't agree and felt that it occurred in the bulk of the solution. I was wondering what your comment would be on the place where the reactions take place.

J. Hoigné • It depends on the pollutant and its concentration. Taking as an example phenol at high pH, the reaction rate constant is 10^9 l/mole sec. This means, for a solution which is millimolar considering phenol, the reaction time will be less than a microsecond, i.e. it must happen in the film. But at much lower concentrations or with compounds such as chloroethylene, it will never happen in the film. It will take seconds (or even days), and in that time ozone will form a homogeneous solution.

Ozone in Water Treatment Processes

H.-P. Klein
Brown Boveri, Zürich, Switzerland

1. Introduction

Today, the quality of raw water is deteriorating due to excessive exploitation and pollution. Special treatment of drinking water as well as of waste water is therefore necessary. Reclamation of waste water for irrigation purposes and ground water reinjection are of vital importance for large areas of the world. As a consequence, the treatment technologies for water have to be adapted to these new requirements. Under these conditions ozonation has gained more importance as a versatile water treatment process. It is especially useful in securing better drinking water even under difficult conditions and in improving waste-water quality. In the industrial sector, the increasing quality demands on ultrapure water for pharmaceutical and electronics production can be fulfilled by application of ozone.

Ozone is an extraordinarily powerful oxidation agent. Since oxidation with ozone yields no other compounds beside the oxidation products and oxygen, ozone is a "clean" oxidation agent. These unique properties explain why ozone has been used in potable water treatment plants for more than 80 years.

So far, ozone has been mainly used in disinfection, in the inactivation of viruses, decoloration, improvement of taste and odour of drinking water and in waste-water treatment. Under today's conditions, additional treatment steps are necessary. Ozonation is introduced in the oxidation of organics. This does not involve complete oxidation but yields products which can be removed by a subsequent treatment step. However, it has been shown that ozone is capable of doing more. It is used as an oxidant in controlling biological contamination and in removing iron, manganese and other heavy metals by precipitation. The latest findings indicate that preozonation enhances microflocculation and increases the filtration rate of rapid sand filters. In all cases the application of ozone helps to decrease reliance on chlorine, which is thought to form by-products harmful to human health.

The aim of this article is to describe some of the latest developments in ozone technology.

2. Advances in Ozone Generation Techniques

Although the commonly used method of ozone generation by silent electrical discharge has been known for more than 130 years, there have recently been some very important advances in this technology which have made ozonation in water treatment more efficient and economic. The introduction of the medium frequency (600 - 1000 Hz) for the electrical power supply has resulted in a space time yield ten times larger than that of the conventional technique with 50 Hz[1]. At the same time, an increase in efficiency and ozone concentration in the carrier gas has been achieved. The resulting specific energy consumption for ozone generation measured

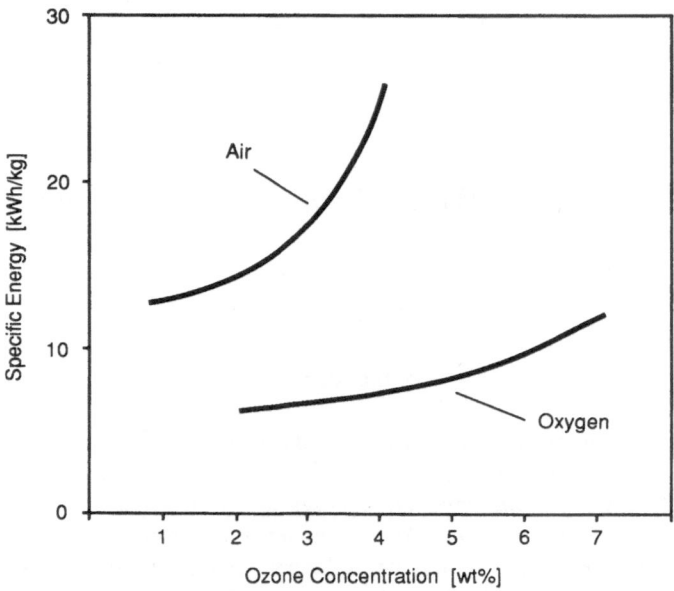

Fig. 1:   Energy consumption of BBC ozone generators as a function of ozone
               concentration with feedgas air or oxygen.

at the ozone generator is less than 12.5 kWh/kg ozone at a concentration of 1.5 wt% (20 g/Nm³)
in air. An ozone concentration of 3 wt% (39 g/Nm³) can be produced with a specific energy
of 15.6 kWh/kg ozone. The advantages of the higher ozone concentration are the smaller gas
handling units and, more importantly, the improved contacting system[2,3] as described below.

Even higher efficiency and concentration can be achieved if oxygen is used as a carrier gas.
The difference in specific energy consumption as a function of ozone concentration for air and
oxygen as carrier gases is shown in Fig. 1. Using oxygen, concentrations of up to 7 wt%
(ca. 103 g/Nm³) are economically attainable. The reduction in size of the ozone generator,
piping, fittings and residual ozone destruction unit are considerable.

The use of oxygen creates no problems for water treatment plants. Today, pure oxygen is
widely used and its large-scale production, storage and transportation are well established.
There are two possibilities for the oxygen supply: oxygen is either obtained from purchased
liquid oxygen (LOX) stored in a tank beside the ozone plant or produced on site. Using a
LOX–tank offers several advantages since its installation, operation and maintenance are very
simple. For fluctuations in water production and therefore ozone demand, it acts as a buffer for
the oxygen supply. Therefore it is a good solution for small and medium-sized plants with
production rates of about 5 to 50 kg/h.

For the on-site production of oxygen, a pressure swing adsorption (PSA) plant is used to
produce up to 500 kg/h oxygen[4] . Only for very large plants with a relatively constant oxygen
demand of a few hundred kilograms per hour may a cryogenic oxygen plant be considered[5].

The use of pure oxygen as the carrier gas has an additional effect. If the carrier gas is air or
a mixture of oxygen and nitrogen, the silent electrical discharge produces, beside ozone, a
small amount of nitric oxides. In the presence of ozone, these are N_2O (nitrous oxide which is
inert), and N_2O_5. In air, the N_2O_5 production is about 1% to 2% of the ozone production [6]. In

a mixture of 95% oxygen and 5% nitrogen (which is typical for oxygen produced by PSA), the N_2O_5 content is still 0.5 to 1% of the ozone portion. Only pure oxygen avoids the formation of nitric oxides. Dissolution in water transforms the N_2O_5 into nitrate. An ozone dose of 2 mg/l increases the nitrate concentration of the water by not more than 0.04 mg/l, certainly a very low value. But, since the influence of the N_2O_5 with ozone on the formation of nitro-compounds is not completely known[7], the use of pure oxygen is preferable.

The new Los Angeles Aquaduct Filtration Plant uses the advantages of an oxygen-fed ozone plant. Since the peak ozone demand is expected to be 150 kg/h, the on-site production of oxygen with a cryogenic plant is considered most economic[8].

More recently, the drinking water treatment plants of Geneva and Zürich have decided to use oxygen as the carrier gas in ozone production. In both plants already under construction, oxygen is fed from an LOX-tank. It has been shown that this results in the lowest treatment costs[9].

All three plants are once-through systems without oxygen recycling. The large fluctuations in production make recycling inefficient.

Another very new development is ozone generation by the Membrel® process[10]. It relies on anodic production of ozone by electrolysis of pure water. The electrolyte is a thin, chemically highly resistant, cation-exchange membrane (e.g. Nafion® from Du Pont). This process produces ozone from water in the water to be treated. Since the specific power consumption is in the order of 60 kWh/kg, this method is only applicable where small quantities of ozone (a few grams per hour) are needed. But it offers enormous advantages for some applications, such as pure and ultrapure water production and conservation[11,12].

3. The Ozone Contacting System

The ozone contacting system, i.e. where the ozone is applied to the water, is very important for the effectiveness of the ozone treatment. This contacting system has the function of finely dispersing the gas in the water, transferring the ozone from the gaseous phase into the water and providing optimum conditions for the subsequent reactions with the various solutes present in the water.

Firstly, the ozone-containing gas is dispersed as small bubbles from which the ozone is transferred into the water. The mass transfer of ozone through the gas-liquid interface can be described by the two-film theory. Diffusion in the liquid film and the bulk of the water determine the mass transfer rate N which is given by:

$$N = \frac{dc_w}{dt} = K_L a (c_{wi} - c_w)$$

(1)

K_L mass transfer coefficient

a   interface area

c_{wi} ozone concentration in water at the interface

c_w ozone concentration in the bulk of the water.

c_{wi} is related to the ozone concentration c_g in the gas phase and the pressure p of the applied gas by Henry's law:

$$c_{wi} = \frac{p}{H} c_g$$

(2)

Henry's constant H, and therefore the ozone solubility in water, shows a very strong temperature dependence. The solubility at 25 °C is about half that at 5 °C, and consequently the mass transfer at 25 °C is slower.

The driving force of the mass transfer is the concentration gradient (c_{wi}-c_w) in the liquid film. The conditions for a high transfer rate and a high efficiency of the contactor are therefore: a high ozone concentration in the feed gas; a large interface area (i.e. small bubbles); a high pressure and a small ozone concentration in the bulk water. The latter is governed by the kinetics of the reactions of ozone with dissolved matter, temperature and pH of the water.

To account for this, eq. (1) has to be modified:

$$N = \frac{dc_w}{dt} = K_L a (c_{wi} - c_w) - k_o c_w [OH^-] - \sum_i k_i c_w [M_i] \qquad (3)$$

k_o rate of self-decomposition of ozone

k_i rate constant of reaction of substance M_i with ozone.

Values of k_i for many organic and inorganic compounds have been determined by Hoigné and co-workers[13].

The most frequently employed types of contactors are: bubble columns with porous diffusers; positive pressure injectors; venturi ejectors and stirred tank reactors. Very often a bubble column with counter-current streams of water and gas is used. Since it utilizes the pressure of the feed gas in ozone production, it needs no additional energy and is therefore the most economic system under normal conditions. Good utilization of the ozone requires column heights of up to 5 m. In order to maintain a minimum residual ozone concentration for a fixed period in e.g. disinfection, the contacting system consists of two or more contacting chambers fed with different ozone doses. Since no contactor gives a complete transfer of ozone to water, the off-gases, which still contain some ozone, are sometimes brought into contact with untreated water in the pre-ozonation stage or in the first contact chamber of the main ozonation stage.

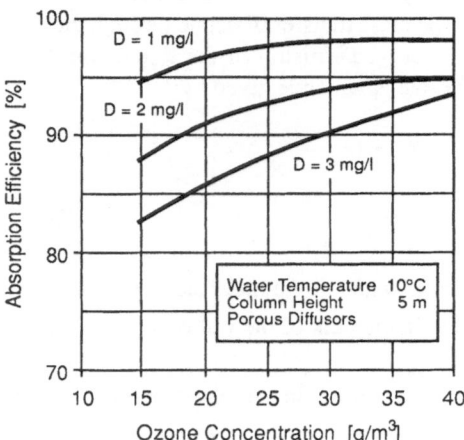

Fig. 2:   Ozone absorption efficiency as a function of ozone concentration at different ozone doses. Porous diffusor system, Dordrecht.

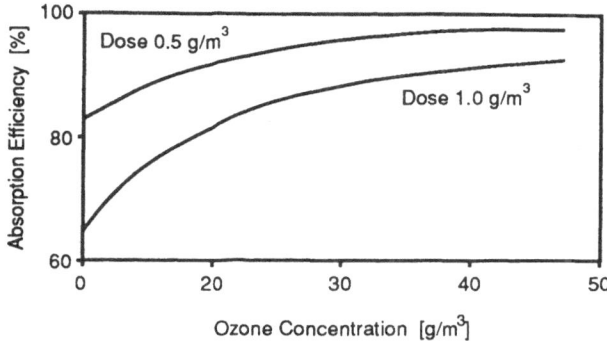

Fig. 3:   Ozone absorption efficiency as a function of ozone concentration at different ozone doses. Radial injector, Horgen.

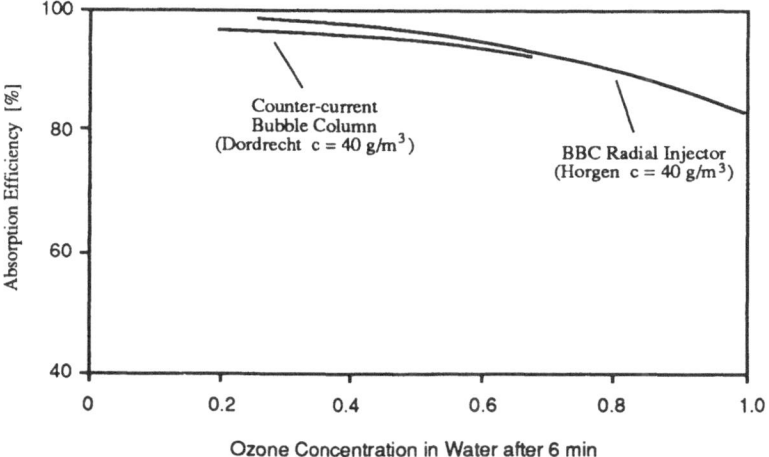

Fig. 4:   Comparison of the ozone absorption efficiency of different contacting systems as a function of the residual ozone concentration.

The positive influence of high ozone concentration in the feed gas on the ozone absorption efficiency is shown in Fig. 2 for various ozone doses. These curves are based on measurements from the GEB waterworks, Dordrecht, The Netherlands, where water from the river Meuse is ozonized in a counter-current bubble column. Fig. 3 shows the same relationship with values obtained in Horgen, Switzerland, where water from Lake Zürich is treated. The ozone transfer device used is a radial injector. A comparison of the results from Dordrecht and Horgen indicates the influence of water quality. Although the contacting system at Horgen should have a higher absorption efficiency at the same dose and concentration because the injector introduces additional mixing energy, it exhibits a smaller one. This is due to the better water quality of Lake Zürich and its lower ozone demand. A better comparison of the two systems which is more or less independent of the water quality is given in Fig. 4. The absorption efficiency is shown as a function of the residual ozone concentration 6 minutes after completing ozonation. It shows that the radial injector has the expected better efficiency for the same residual ozone concentration.

Even higher absorption efficiencies can be achieved when ozone is produced from oxygen in concentrations up to 7% by weight. In the new Los Angeles Aquaduct Filtration Plant, the use of oxygen with an ozone concentration of 6% results in an absorption efficiency up to 95%. As a consequence, the design capacity of the ozone production facility could be reduced by 5%, leading to corresponding cost savings.

4. Ozone Applications

4.1. Drinking Water Treatment With Ozone

Since the beginning of the century, ozone has been used predominantly as disinfectant, usually applied after filtration. During the last two decades ozone has been employed for several other purposes at different stages in the water treatment process. In the general treatment scheme in Fig. 5, the points at which ozone can be applied are indicated (pre-ozonation, main ozonation and post-ozonation)[14,15].

4.2. Pre-ozonation

Pre-ozonation is primarily applied in: removing iron and manganese by precipitation; partial improvement of the organoleptic properties i.e. colour, taste and odour; and elimination of algae and reduction of the trihalomethane (THM) formation potential. Pre-ozonation replaces the pre-chlorination step and thus no halogenated organic compounds are formed at that stage[14]. Moreover, many surface water treatment plants have an additional ozone production capacity installed in case of environmental accident when, e.g. phenolic compounds could spoil the water[16].

Ozone has recently been used to support microflocculation[17]. This term comprises: the formation of particles; changes in the colloidal character of the dissolved matter; and reduction in turbidity and improvement in the subsequent sedimentation and filtration stages. The most accepted explanation of this phenomenon is as follows:

Finely dispersed minerals, such as quartz, clay and diatomaceous earth particles adsorb organic colloidal substances, e.g. humic acids and fulvic acids, which are present in water, especially in surface water. The adsorbate stabilizes the finely dispersed particles by hindering their coagulation as a result of repulsive forces between the colloidal particles[18,19]. Ozone partially oxidizes the colloids, thus reducing the repulsive forces between the particles. Polar and chelating groups are formed which induce coagulation. This mechanism explains why an excessive dose of ozone, which results in excessive oxidation of small water soluble organic molecules, has an adverse effect i.e. hinders the flocculation. It has been shown that if ozone doses from 0.1. to 0.4 mg/mg DOC are applied together with the classical flocculation agents (Al^{3+} or Fe^{3+} salts), optimum flocculation is achieved. Compared with conventional flocculation, the addition of ozone can lead to savings of up to 50% of the flocculation agent. On the other hand, if removal of DOC as well as of iron and manganese is the objective of ozonation, the ozone dose has to be 0.5 to 1 mg/mg DOC[15,20]. In this case, flocculation can only

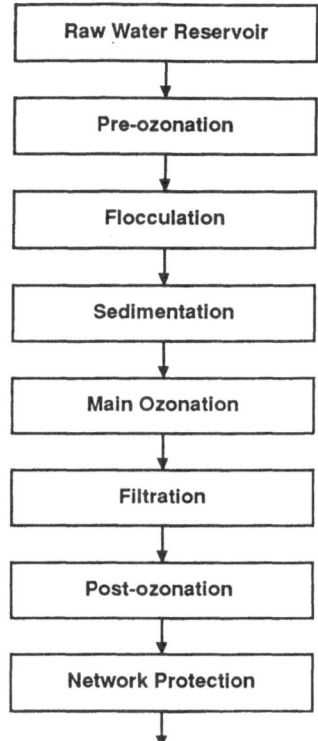

Fig. 5:   Points of application of ozone in a general water treatment scheme.

be achieved if the same amount of flocculation agent is added as in flocculation without ozone. Pre-ozonation for microflocculation and pre-ozonation for oxidation have therefore converse effects.

The new Los Angeles Aquaduct Filtration Plant is a typical example of microflocculation induced by ozone[21]. Over several years, pilot plant experiments with different pre-treatment processes before filtration have been carried out. The results indicate that the use of ozone has many advantages compared to other pre-treatment processes:

- The filtration rate is increased by more than 50%; this allows a 30% reduction in the number of filters.
- The time for flocculation is reduced by 50%. Hence, only half the number of flocculation basins are needed.
- The period between two backwashings is extended. As a consequence, the size of the treatment facilities for the backwash water can be reduced and the amount of sludge decreased.
- Ozone supports microflocculation and reduces reliance on other chemical additives (coagulants) by one third.
- Further recognized benefits are the reduced chlorine demand, the improvement of organoleptic properties (i.e. taste and smell), reduction of the THM formation potential by 50% and lower turbidity values.

Similar results have been obtained in water treatment plants in Europe[15].

4.3. Main Ozonation

In the main ozonation stage, the appropriate dosage of ozone is used for intermediate disinfection and oxidation of organic substances. These substances comprise natural and synthetic organic matter. The complex chemistry and kinetics of reactions of ozone with different compounds in aqueous solution have been studied extensively by Hoigné and his group[13]. In practice, the organic matter is not converted to CO_2 and H_2O by ozonation, but only to oxygen containing organic compounds like carboxylic acids, aldehydes or ketons. Complete oxidation to CO_2 turns out to be uneconomic because, beside the time needed for further oxidation, the ozone demand rises drastically.

Much work has been devoted to the investigation and identification of ozone by-products[22], their impact on water quality with respect to human health and their eventual removal from water if this appears to be necessary. The most important finding of these studies is that ozonation alone generally cannot guarantee healthy water. Rather, it has been found that some by-products formed are still toxic or even mutagenic. Ozone appears either to increase or decrease mutagenicity depending on treatment conditions and the nature of the organic matter present in the raw water[22,23]. Even the trihalomethane (THM) formation potential of the water is often increased due to the fact that ozonation forms smaller organic molecules which result, after chlorination, in organic molecules containing more chlorine. However, in oxidation with ozone, the by-products formed are generally more accessible to removal by biological treatment than the untreated molecules or those formed by chlorination. Even refactory (non-biodegradable) compounds, such as some pesticides, can be transformed into biodegradable compounds[24]. Ozonation prior to filtration can promote biological activity of the filter which then removes the oxidation products, including any THM precursors.

In all cases, the main ozonation has to be followed by filtration. Filtering through granulated activated carbon (GAC) removes the oxidized organics. Under such conditions, the GAC develops a high degree of biological activity[14]. The advantage of the biologically active GAC is the extension of the useful lifetime of the carbon due to continuous regeneration by the biochemical degradation of the adsorbed organic material. It has to be emphasized that the biological GAC filtration process works satisfactorily only after ozonation.

A similar biological activity may develop in slow sand filters or diatomaceus earth.

4.4. Disinfection

Disinfection, the most important stage in the purification process of water, was the first and, for a long time, the predominant application of ozone in water treatment. The disinfection power of ozone derives from its oxidation potential. Ozone attacks and destroys the cell membrane of bacteria. The resulting cell lysis makes a reactivation of the bacteria impossible. Because of this direct attack on the cell membrane by ozone, the killing of bacteria and inactivation of viruses is very fast.

To guarantee the necessary degree of disinfection, a minimum ozone content has to be maintained for a certain period of time in the water. These data are specific for every type of microorganism and depend on the water temperature, a lower temperature needing longer periods.

Due to its rapid decay into oxygen, ozone lacks the residual effect which is necessary for the protection of the water in the distribution network system. Moreover, the reaction products of ozonation with natural organic substances are low molecular-weight oxygenated molecules that are usually more easily biodegradable than the original substances. These oxidation products may promote biological growth in the distribution network instead of hindering it. This drawback is overcome by adding a small residual dose of chlorine or better, chlorine dioxide, to the ozonated water just before leaving the plant. Since the water to be chlorinated contains no living microorganisms, the necessary applied dose of the residual disinfectant is very small (in the order of 0.1 mg/l).

It should be mentioned that ozone treatment provides a good basis for ground water reinjection, where surface water has to be brought up to drinking water quality before being injected into the aquifer.

For the decontamination of oil-polluted ground water, the city waterworks of Karlsruhe are using ozone treatment of the polluted water which is subsequently fed into the ground water[25].

5. Waste Water

The two main applications of ozone in waste-water treatment are disinfection and, especially in industrial waste, oxidation.

5.1. Disinfection

Municipal effluents discharged into the environment are a source of pathogenic contamination. In areas where the natural safeguards, such as dilution and residence time before use, are not sufficient, disinfection is necessary. In North American and in arid regions such as the Middle East where extensive water recycling and reclamation are usual, waste-water disinfection is in general practice. Chlorine, commonly used as disinfectant in the past, effects aquatic life since it forms chlorinated organic compounds which are potentially toxic. As they are hardly biodegradable, they accumulate.

Ozonation provides an alternative[26] without these disadvantages. Ozone itself is toxic to aquatic life, but it decomposes rapidly to oxygen after its application before discharge of the treated water. In some instances, toxic, potentially mutagenic and/or carcinogenic compounds have been found as a result of reactions with organic matter. But such compounds are transformed by the dissolved oxygen or by biodegradation. Little is yet known about these by-products but it is thought that ozonation presents less of a danger to the environment than chlorination.

Due to the higher concentration of organics and inorganics in waste water, more ozone is required for disinfection than in the case of drinking water. The ozone dose needed to achieve the necessary reduction in bacteria counts and to meet the respective disinfection standard 2.2, (70 or 200 total coliforms per 100 ml) depends on the quality of the treated water. It ranges from about 5 mg/l for a rotating biological contactor effluent to more than 30 mg/l for highly contaminated treated waste water. Since the effluent quality varies from plant to plant, the ozone dose necessary to reach the desired disinfection standard has to be determined by on-site pilot plant tests.

5.2. Oxidation

Since ozone is such a strong oxidation agent and reacts very selectively (e.g. with double bonds), ozonation shows promise in solving industrial waste-water treatment problems. The aim of ozonation is, in most cases, not complete oxidation (mineralization), but rather oxidation that yields non-toxic, biodegradable compounds[22] which can be removed afterwards by a cheaper process. With stricter legislation concerning the purification of industrial waste water, ozonation has become an important means of fulfilling these requirements.

A recent application of ozone in industrial waste-water treatment is the oxidation of heavily toxic cyanide in galvanic effluents to non-toxic cyanate.

6. Process Water Treatment

6.1. Cooling Water

During the last ten years, many reports on the use of ozone as an alternative biocide for cooling-tower water systems have been published. The results are generally very promising. Ozone has been found to be very efficient in controlling microbiological growth and may even prevent corrosion and scaling. It seems that water treatment with ozone exclusively, i.e. without any additional chemicals, is usually sufficient. But information is contradictory particularly with respect to the ozone dose necessary (ranging from 0.03 to 0.75 mg/l and more)[27,28], its point of injection and its effectiveness. Fortunately, the latest results confirm the effectiveness of small doses of 0.05 to 0.1 mg/l [29]. Nevertheless, ozonation has yet to be applied on a large scale, e.g. in power plant cooling-tower water systems. It is likely that the economical and environmental advantages of this method compared with chlorine or other chemicals will lead to a break-through in the near future.

6.2. High Purity Water

Modern ultrapure water systems are mostly designed as closed loop systems in which the water is pumped from a storage tank through one or more recirculation loops to the production facilities. Water which is not consumed is returned to the storage tank. Since non-flowing water in the distribution lines allows the growth of bacteria, the water has to be circulated continuously, even when production is shut down. Disinfection is still necessary, at least periodically. Ozone is the most effective and suitable disinfectant for this purpose even if it is applied in very small concentrations. Since it disintegrates rapidly (its half-life is about 20 minutes), excessive concentration cannot occur even with a constant dosage of ozone. If pure water has to be absolutely free of ozone, any residue can be easily decomposed to oxygen by means of ultraviolet irradiation[30].

The disinfection of ultrapure water systems has benefited from the development of Membrel technology, which produces ozone electrolytically in the water to be treated[11,12]. This allows disinfection without introducing any potential contaminants. In contrast to ultraviolet light, ozone still has a residual effect for a limited period.

Another possible application of ozone and the Membrel process may be disinfecting water before it passes through ultrafiltration or reverse osmosis modules[31]. The efficiency of ozone in controlling the biofouling of the membrane has been proved, but most membrane types are not chemically stable in ozone and therefore the ozone has to be destroyed before the water reaches the membrane.

7. Summary

Ozone is a very versatile oxidation agent which can help to solve a lot of different problems such as disinfection, removal of organics, improvement of organoleptic properties etc. in all kinds of water treatment. As a substitute for chlorination, ozonation reduces the level of harmful chlorinated organic compounds present in the water. To obtain the maximum effect of ozonation in each application, an exact definition of its function has to be worked out. This is important in order to determine the appropriate point of introduction, the applied dosage and the necessary pre- and post-treatment. Two or even three-stage ozonation, which takes these aspects into consideration, is gaining in popularity.

References

1   Gaia, F. and Menth, A., New High-Power Ozone Generators and their Application in Industry. Ozone Sci. Eng. **4** (1982) 207-214.

2   Klein, H.-P. and Steiner, H.P., Ozon in der Trinkwasseraufbereitung. Brunnenbau, Bau von Wasserwerken, Rohrleitungsbau (bbr) **36** (1985) 186-193.

3   Fischer, M., Klein, H.-P., Liechti, P.A. and Dyer-Smith, P., Technical and economical advantages of producing and applying ozone at high concentrations. Ozone Sci. Eng. **9** (1987) 93-108.

4   Riquarts, H.-P. and Leitgeb, P., Gastrennung mit Druckwechsel-Adsorptionsanlagen. Chem.-Ing.-Techn. **57** (1985) 843-849.

5   Dyer-Smith, P. and Jaisli, E., Economical large-scale production of ozone and its practical application. Brown Boveri Review **72** (1985) 372-375.

6   Kogelschatz, U. and Bässler, P., Quantitative Angaben zum N_2O- und N_2O_5-Ausstoss moderner Hochleistungsozonisatoren bei der Ozonerzeugung aus Luft. bbr **36** (1985) 453-456.

7   Becke, Ch. and Maier, D., Herkunft von Trichlornitromethan in Trinkwasser. Vom Wasser **62** (1984) 125-135.

8   Becke, Ch. and Maier, D., Die Rolle der Stickoxide bei der Wasseraufbereitung mit Ozon. Vom Wasser **59** (1982) 269-276.

9   Geering, F., Experiences with ozone treatment of water in Switzerland. Proc. 8th Ozone World Congress, Sept. 1987, Zürich. International Ozone Association (IOA). (Unionsverlag, Zürich) 1987, Vol. 1, pp. B59 – 75.

10   Stucki, S., Theis, G., Kötz, R., Devantay, H. and Christen, H.J., In situ production of ozone in water using a Membrel electrolyzer. J. Electrochem. Soc.: Electrochem. Sci. Techn. (1985) 367-371.

11   Vogel, L. and Klein, H.-P., Ozone generation by means of Membrel electrolysis: a process for treating ultra-pure water. Brown Boveri Review **73** (1986) 451-456.

12   Setz, W., Erfahrung mit einer Verfahrenskombination "Umkehrosmose/Kontinuierlicher Ionenaustauscher" zur Herstellung von Wasser für pharmazeutische Zwecke. Pharm. Ind. **47** (1985) 3-11.

13   Hoigné, J. and Bader, H., Rate constants of reactions of ozone with organic and inorganic compounds in water. Part I: Non-dissociating organic compounds. Water Res. **17** (1983) 173-183. Part II: Dissociating organic compounds. Water Res. **17** (1983) 1985-1995. Part III: Inorganic compounds and radicals. Water Res. **19** (1985) 993-1004.

14   Sontheimer, H., Heilker, E., Jekel, M.R., Nolte, H. and Vollmer F.A., The Müllheim process. J. AWWA **70** (1978) 393-396.

15   Rice, R.G., Rationales for multiple stage ozonation in drinking water treatment plants. Ozone Sci. Eng. **9** (1987) 37-62.

16   Schalekamp, M., Raw water quality and water treatment. Zbl. Bakt. Hyg. I Abt. Orig. **B 172** (1980) 156-180.

17   Maier, D., Mikroflockung durch Ozon. Paper presented at the Conf. on "Oxidationsverfahren in der Trinkwasseraufbereitung", Karlsruhe, FRG, Sept. 11-13, 1978. W. Kühn, H. Sontheimer, eds. (Karlsruhe) 1979, pp. 417-441.

18 Jekel, M., Ozone for microflocculation. DVGW Schriften Wasser **42** (1985) 137-143.

19 Jekel, M., *Einfluss von natürlichen organischen Stoffen auf die Kolloidstabilität von praktikularen Substanzen.* Bd. 20 der Veröffentlichung des Bereichs Wasserchemie Karlsruhe. H. Sontheimer, ed. (ZfGW-Verlag, Frankfurt) 1982, pp. 287-300.

20 Sontheimer, H., Trends in ozonation (Roundtable Discussion). J.AWWA (1985) 30.

21 Stolarik, G.F., Ozonation and the direct filtration process. Paper presented at American Water Works Association, California - Nevada section. Fall Conf. Oct. 28-30, 1981, Palm Springs, CA.

22 Glaze, W.H., Reaction products of ozone: A review. Environmental Health Perspectives **69** (1986) 151-157.

23 van Hoof, F., Janssens, J.G. and Van Dijck, H., Formation of mutagenic activity during surface water preozonation and its removal in drinking water treatment. Chemosphere **14** (1985) 501-509.

24 Medley, D.R. and Stover, E.L., Effects of ozone on the biodegradability of biorefractory pollutants. J. WPCF **55** (1983) 489-494.

25 Nagel, G., Kühn, W., Werner, P. and Sontheimer H., Grundwassersanierung durch Infiltration von ozontem Wasser. gwf-Wasser/Abwasser **123** (1982) 399-407.

26 Design Manual: Municipial Wastewater Disinfection EQP-Report EPPA/625/1-86/021 Oct. 1986, Chapter 6, pp. 97-155. US Environmental Protection Agency, ed. Office of Research and Development. Water Engineering Research Laboratory Center for Environm. Res. Information. Cincinnati, OH 45 268.

27 Merrill, D.T., Parker, D.S. and Drago, J.A., Evaluation of ozone treatment in cooling towers. Proc. 35th Industrial Waste Conf. 1980, pp. 307-315.

28 Siegrist, H.W., Tuttle, D.G. and Majumdar, S.B., Technical and economic considerations of biocide system options for cooling water systems; A review. Proc. 2nd Intern. Symp. on Ozone Technology held at Montreal, Canada, May 11-14, 1975. R.G. Rice, P. Pichet and M.-A. Vincent, eds. pp. 632-649.

29 Leitzke, O. and Greiner G., Wasserbehandlung in Rückkühlkreisläufen mit Ozon. Vom Wasser **67** (1986) 49-58.

30 Baker, R. and Taylor, F.M., Oxidation of 2 propanol in dilute aqueous solution by UV/ozone. Proc. Intern. Conf.: The Role of Ozone in Water and Waste Water Treatment. R. Perry and A.E. McIntyre, eds. London 13-14 Nov. 1985. pp. 106-116.

31 Lozier, J.C. and Sierka, R.A., Using ozone and ultrasound to reduce RO membrane fouling. J. AWWA **77** (1985) 60-65.

Discussion

Chairman: E. Merz
Brown Boveri, Zürich, Switzerland

J. Stankovic • There is controversy about what should be done with excess ozone from an ozonation plant, is recycling it to the preozonation stage or destroying it more economic? What do you suggest is the best solution to the "off-ozone" problem? And what contacting equipment do you use in the preozonation stage?

H.P. Klein • About 5 to 10% of ozone is left in the off-gas from the contacting system. Careful calculations have shown that it is not economic to recycle this ozone because the complete volume of the carrier gas has to be pressurized in order to bring it into contact with the water in the preozonation stage. It is more economic to destroy the ozone in the off-gas and to increase instead the ozone production capacity.

J. Stankovic • Yes, but you mentioned contacting efficiencies of 80 to 85% in the 5 meter bubble column. In my opinion these efficiencies are a bit low.

H.P. Klein • The contacting efficiency depends on the ozone dose and the ozone demand of the water. The curves you refer to include high doses, which normally are not applied in waterworks. Under normal operation conditions, i.e. if the dose is adjusted to a residual ozone concentration of 0.4 or 0.6 mg/l, absorption efficiencies of more than 90% will be achieved. The second question was what unit would be used for preozonation to introduce the ozone. If the water quality is good, as for instance in Lake Zürich, porous diffusers can be used. If the water has a high manganese, iron etc. content, an injector will be needed.

J. Stankovic • Do you use constant-speed blowers or variable-speed blowers to bring ozone in contact with water?

H.P. Klein • We adjust the blowers to operate with a constant concentration in the gas.

Disinfection with UV-Radiation

Clemens von Sonntag

Max-Planck-Institut, Mühlheim/Ruhr, West-Germany

1. Introduction

There are many well-advanced techniques for the disinfection of surface water producing the hygienic quality required for drinking water. Most of these techniques (e.g. gassing with chlorine, chlorine dioxide, or ozone) make use of oxidation of the organic matter in water. This does not exclusively degrade the object of interest, i.e. the microorganisms so that, in competition with disinfection, much more material is degraded than actually required. Irrespective of the oxidant, the degradation products are numerous and there is increasing concern about the adverse effects of some of these products on humans and the environment. Notable among these are the chlorinated compounds produced when water is disinfected with chlorine. In addition, chlorine dioxide and chlorine give an unpleasant flavour to drinking water when still present at the consumer's end (some ozone-derived products have this property as well). The public has become increasingly reluctant to accept such water. Furthermore, at certain times during the summer, some surface water contains so much organic material that disinfection with chlorine dioxide can only be achieved if the maximum permissible dosage is added. Any new legislation reducing this limit would cause considerable problems for reliable disinfection. For this reason, alternatives must be sought. A promising alternative is disinfection by means of UV-radiation.

Disinfection of water by UV-radiation is by no means a new idea. In fact, it was among the very first of the processes still in present use to be introduced. However, cost and technical pro-

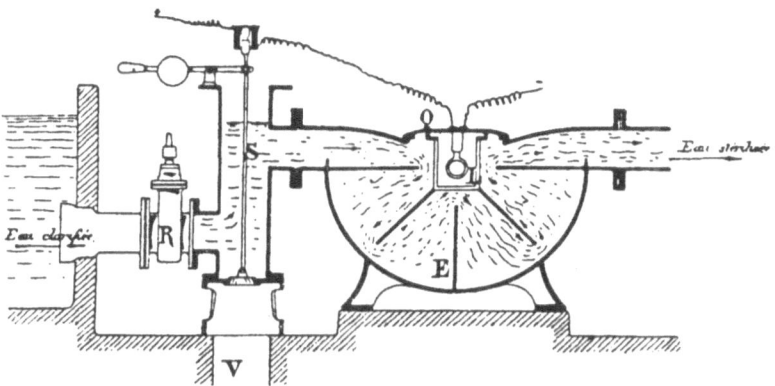

Fig. 1:   A 1910 design of a UV-disinfection plant[1].

blems have meant that other processes were able to compete successfully with UV-irradiation. Already, in 1910, equipment was designed which allowed disinfection of water on a large scale (Fig. 1)[1]. The Hg-lamps then used were of a comparatively simple design (Fig. 2)[2]. Nevertheless, the scale-up was successful, and a pilot plant was installed near Paris. It was large enough to supply UV-disinfected drinking water for a community of as many as 20,000 people (Fig. 3)[3].

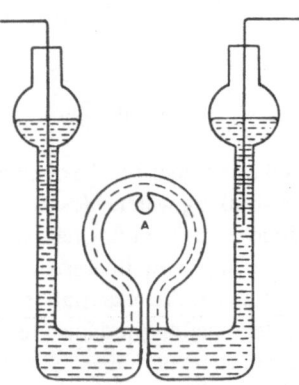

Fig. 2:   Hg-arc for UV-disinfection. A 1910 construction[2].

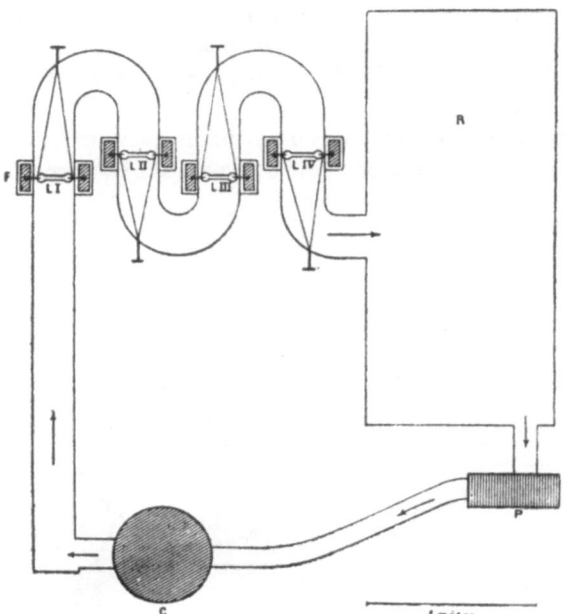

Fig. 3:   A scale-up of a UV-disinfection plant (1910) to supply drinking water for a community of 20,000 people[3].

Today, there are many small and medium-sized plants which make use of this technique. They mainly use rather clear water from mountainous regions. In Germany, the "Bundesminsterium für Forschung und Technologie" has now launched a program to investigate whether or not this technology can also be implemented under less favourable conditions, such as for water from open reservoirs and lakes.

It is interesting to note that one of the major issues of concern in these early publications was the effective and economic utilization of light. The flow of water in the reactor was channelled or deflected by baffles in various ways as can be seen in Fig. 1 and 3. Although at present hydrodynamics is still a major problem, I will not dwell on this point but rather deal with some biological and photochemical aspects of water disinfection by UV-radiation.

Misconceptions still abound about primary UV photochemistry in water, especially concerning the purported participation of dissolved molecular oxygen (cf. Ref. 4). It is important for the proper development of UV-water-disinfection technology that these erroneous ideas be clarified.

2. Rationale of UV-Disinfection

Disinfection of surface water for drinking purposes aims to reduce the number of microorganisms to a level acceptable to present hygienic standards. It must also be guaranteed that, during the usual residence time of the disinfected water in the pipeline network of the water supply system, no or only negligible regrowth of the microorganism population occurs.

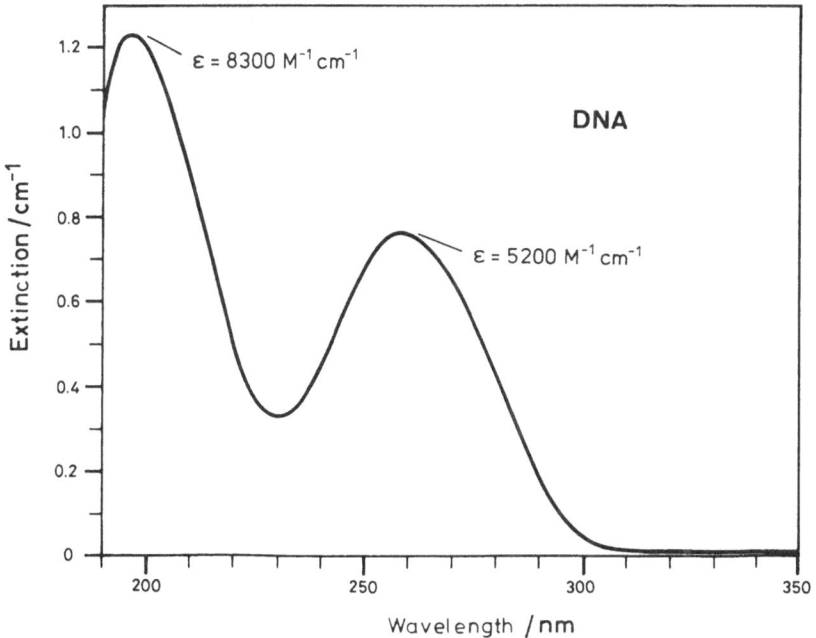

Fig. 4: UV-absorption spectrum of DNA (cf. Ref. 5).

It is sufficient if the microorganisms originally present are prevented from multiplying. A microorganism which continues to metabolize but which is unable to divide may be regarded as dead in this context (reproductive cell death). Reproductive cell death requires smaller light doses than vegetative cell death.

In microorganisms, UV light is largely absorbed by the nucleic acids DNA and RNA (for some strongly coloured bacteria see below). The UV-spectrum of DNA is shown in Fig. 4.

DNA (in some viruses, RNA) is the key molecule which stores all the genetic information (genome) of the microorganism. Its full genome must be passed on to its daughter cells (daughter viruses) to ensure propagation. In Fig. 5, a section of the DNA macromolecule is shown.

The UV-absorption of DNA is caused by the nucleobases, i.e. by the purines, adenine and guanine; and the pyrimidines, thymine and cytosine (in RNA thymine is replaced by uracil which lacks the methyl group at C 5). The sugar-phosphate backbone starts to absorb only at much shorter wavelengths ($\lambda \leq 210$ nm). This fact is of little concern in the present context, because, for disinfection, light of $\lambda \geq 220$-240 nm is usually employed.

As discussed later in detail (see section 5), the damage inflicted on the DNA by UV-radiation is mainly brought about by photochemical reactions of the pyrimidines, thymine and cytosine. The reaction products prevent the duplication of the nucleic acids and hence the reproduction (propagation) of the microorganisms.

While the oxidative degradation of the microorganisms by other water disinfection procedures[5] is not specific, and lipids and proteins as well as nucleic acids are degraded more or less equally readily, UV-radiation mainly degrades DNA which represents only about one percent of the organic matter in cells. Thus UV-radiation causes minimal chemical changes in

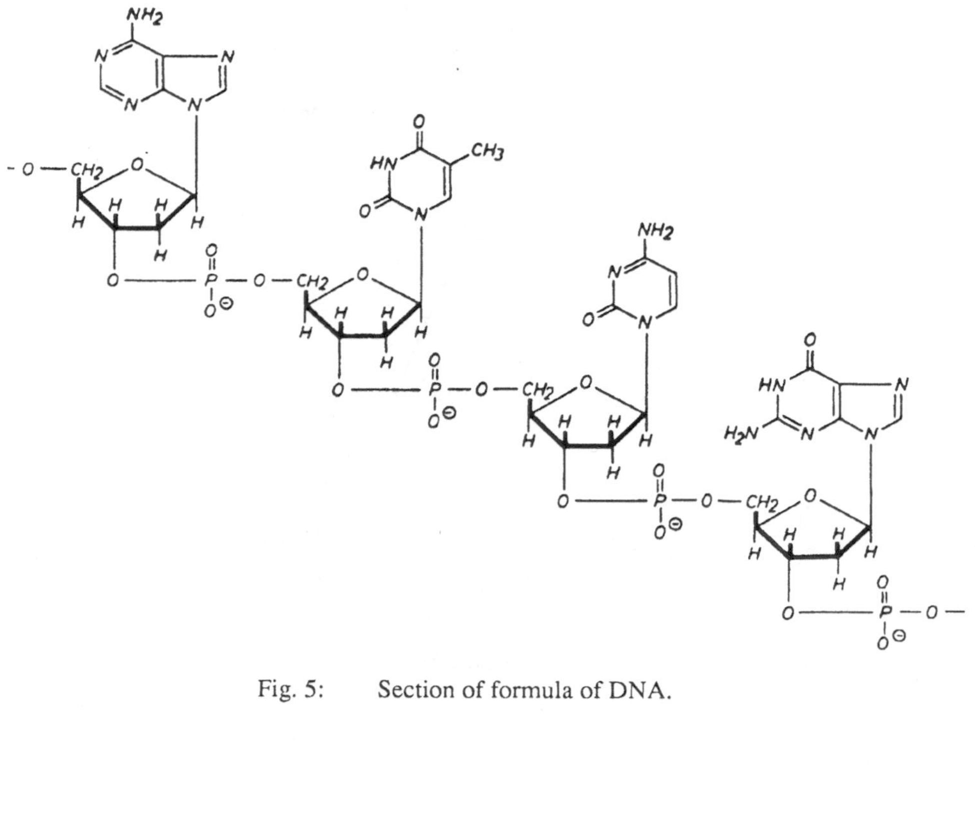

Fig. 5:   Section of formula of DNA.

a disinfection reaction. It's not yet clear whether this also holds for other organic and inorganic material present in the water to be disinfected (see section 7).

3. UV-Light Sources

At present, two different kinds of UV-light sources are used in disinfection, namely the low-pressure mercury (Hg) arc and the high-pressure Hg-arc. The low-pressure Hg-arc mainly emits one line at 254 nm (the 185 nm line is usually absorbed by the doped quartz of such lamps), the remaining long-wavelength light being of the order of 10% (Fig.6). The high-pressure Hg-arc, however, exhibits a large number of lines which are superimposed on a broad continuum (Fig. 7).

On comparing the emissions of these two light sources with the absorption spectrum of our target, DNA (Fig. 4), one is tempted to consider the low-pressure Hg-arc to be more cost-effective, because little light (that means also less electrical energy) is lost where DNA does not absorb $\lambda > 300$ nm). Furthermore for short wavelengths, a quartz cutting off at $\lambda = 220$–240 nm appears to be generally used for disinfection purposes. Much valuable energy is lost when the high-pressure Hg-arc is used (Fig. 7). There are, however, other factors besides the price of electricity which determine the overall cost per m³ of water to be disinfected and much remains to be done in the development of low-priced and long-lived light sources. At present, which kind of light source will eventually be used cannot be foreseen. In later sections, use is made of the different emission characteristics of two light sources in order to estimate the possible contributions of photochemical side reactions involving organic and inorganic material present in the surface water to be disinfected (see section 7).

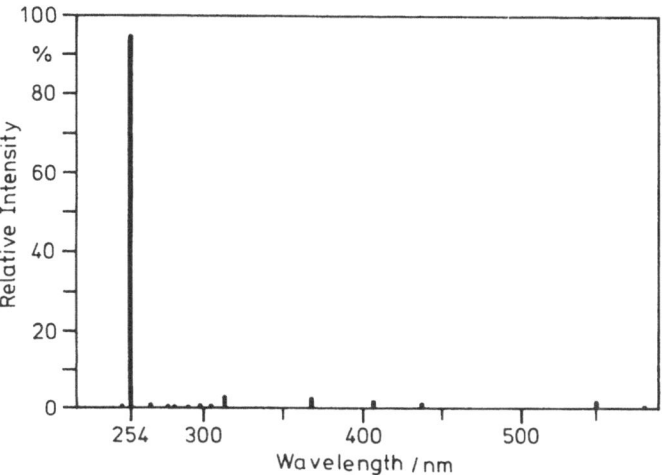

Fig. 6:   Emission spectrum of the low-pressure Hg-arc.

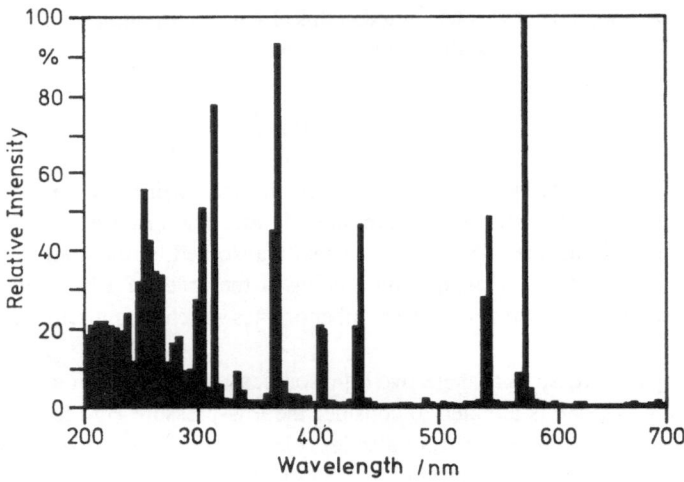

Fig. 7:   Emission spectrum of the high-pressure Hg-arc.

4. Biological Inactivation and Quantum Yields

As mentioned above, a living cell or a virus when subjected to UV-radiation becomes inactivated. An outline of our present state of knowledge about inactivation is given below. First we deal with quantitative aspects of the inactivation process. These must be related to other effects of UV-radiation, i.e. to the chemical changes of other material which is present in the water (see section 7).

The inactivation of biological material follows two slightly different patterns shown in Fig. 8 and 9. In Fig. 8 the number of viable cells (viruses) divided by the number of such species originally present (n/n_0) (termed survival) is plotted on a logarithmic scale *vs.* the UV-fluence, which is proportional to the UV-dose absorbed by the microorganisms (see section 6). The degree of inactivation is represented by a straight line over several orders of magnitude of n/n_0. Such inactivation curves are observed if no repair processes (see section 6) occur. In Fig. 9, the same quantity is plotted, but at the lower fluence range a shoulder is observed, and only at the higher fluences does the curve become linear. The shoulder is usually due to enzymatic repair processes within the microorganism. It is evident that only the linear sections of these two types of curves can be compared with one another, and the shoulder at the beginning of the curve in Fig. 9 is therefore often neglected when inactivation data are given. The L_{37} (often called D_0) fluence required to reduce the viable fraction to 37% (1/e) is inversely proportional to the slope of the linear part of the curve (in many cases the shoulder may be quite pronounced and D_0 values are inadequate to compare the sensitivity of two organisms). In such inactivation studies, the UV-exposure suffered by the microorganisms is usually given in fluence units (J m^{-2}). This unit is different from the unit of the quantum yield (Φ) which photochemists prefer to use. The quantum yield is a more straightforward quantity. It is defined as the number of molecules formed per number of photons absorbed. Since 1 einstein represents 1 mole of quanta, the dimension of the quantum yield is mole einstein^{-1}.

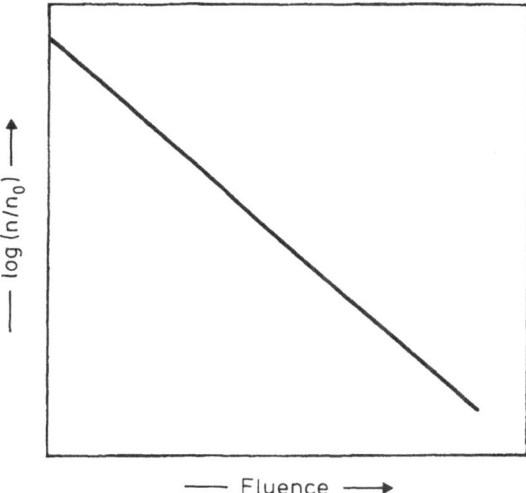

Fig. 8: UV-inactivation of microorganisms not possessing a repair system. Log(survival) (log n/n_0) is plotted against the fluence (J m^{-2}).

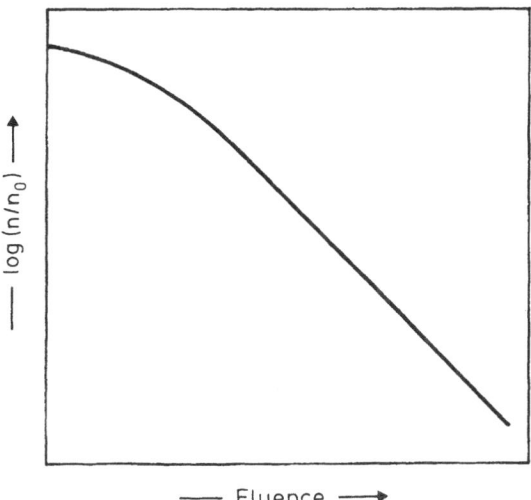

Fig. 9: UV-inactivation of microorganisms capable of repair. Log(survival) (log n/n_0) is plotted against the fluence (J m^{-2}).

The quantum yield can also be given in terms of a ratio of cross sections:

$$\Phi \;=\; \sigma_p / \sigma_a \tag{1}$$

where σ_p is the conversion cross-section and σ_a the absorption cross-section. The conversion cross-section is related to the number of product molecules formed (Δn_p) when n_s molecules of the parent are exposed to a small photon fluence ΔH (dimension: number of photons divided by the area of the cross-section of the incident beam):

$$\Delta n_p \;=\; \sigma_p \, n_s \, \Delta H \;. \tag{2}$$

This expression is valid under conditions where there is negligible absorption by the substrate within the illuminated medium (this holds for most photobiological systems and also for dilute nitrate solutions, see section 7).

If several products are formed along different reaction channels $S \rightarrow \alpha_1 P_1,\ \alpha_2 P_2, \dots$, ($\alpha_i$ = stoichiometric factors), then:

$$\Delta n_s \;=\; -\sum \Delta n_{P_i} / \alpha_i \tag{3}$$

i.e.

$$\Delta n_s \;=\; -n_s \, \Delta H \sum \sigma_{P_i} / \alpha_i \;. \tag{4}$$

Upon integration,

$$n_s \;=\; (n_s)_0 \times e^{-\Sigma \sigma_i / \alpha_i H} \tag{5}$$

where $(n_s)_0$ is the number of parent molecules at the start of the illumination, and H the integral fluence. In the special case where none of the products absorb light at the photolysis wavelength,

$$n_s / (n_s)_0 \;=\; A / A_0 \tag{6}$$

(A = absorptivity) and one may determine $\sum \sigma_i / \alpha_i$ from the slope of the plot of $\ln(A/A_0)$ vs. H. In the case where only one single product is formed in a 1:1 stoichiometry, the corresponding photochemical cross-section is accessible from the absorptivity measurements. The fluence H equals the light intensity (I) multiplied by the irradiation time (t):

$$H \;=\; I \times t \;. \tag{7}$$

The light intensity can be determined by actinometry. It is thus possible to determine quite accurately the conversion cross-section without knowing the extinction coefficient of the species involved. For biological systems, reliable absorption spectra are often not known, but they are easy to measure in the case of model compounds for which quantum yields can then be given.

Biologists prefer the unit J m^{-2} for the fluence. To relate it to the quantum yield one must keep in mind that the energy content of 1 mole of quanta (1 einstein) depends on the wavelength of the quanta ($E = h\nu = hc/\lambda$; for 254 nm quanta 1 einstein $= 4.71 \times 10^5$ J). The dimension of σ_p is $m^2 J^{-1}$, and that of σ_a is $m^2 mole^{-1}$. The decadic extinction coefficient ε usually used has the dimension of $dm^3 mole^{-1} cm^{-1}$. Hence σ_p and ε have the same physical dimension. In numerical terms, $|\sigma_a| = 2.3 \times 0.1|\varepsilon|$ i.e. the quantity $\sigma_a = 0.23 |\varepsilon| m^2 mole^{-1}$. Since $\sigma_p = 1/H_{37}$ ($1/D_0$) eq. (8) holds for 254 nm quanta when the fluence is given in J m^{-2}:

$$H_{37} = \frac{4.71 \times 10^5}{\Phi(254) \times 0.23 \times \varepsilon(254)} \; [\, J\, m^{-2}]\quad. \tag{8}$$

Below we will make use of this equation when we compare biological inactivation and nitrite formation in surface waters disinfected by UV-radiation (section 7).

5. The Photochemistry of Nucleic Acids

The photochemistry of nucleic acids and their model compounds has met with considerable interest in the past, and a number of excellent reviews are available[6-14]. It will therefore suffice here to give a very brief account of the nature of the major photochemical products and to discuss in the following section some of the biological consequences.

The amount of absorbed UV light by the purines and the pyrimidines is roughly equal (they have similar extinction coefficients), but only the latter give rise to photo-products with reasonably high quantum yields. Most prominent is the formation of the cyclobutane-type dimers Pyr< >Pyr (**1-3**):

1

2

3

Such dimers can be formed when two pyrimidines are adjacent to one another in the same strand (Pyr .. Pyr).

$$Pyr..Pyr \underset{h\nu}{\overset{h\nu}{\rightleftharpoons}} Pyr < > Pyr \quad . \tag{9/10}$$

Photochemical dimer formation (reaction 9) is reversible (reaction 10). In the Pyr< > Pyr dimer, the 5,6-double bond of the pyrimidines is lost. Hence, it no longer absorbs strongly in the 260 nm region (cf. DNA spectrum, Figure 4). It absorbs significantly, however, at shorter wavelengths where the dimer can be split, thus allowing the integrity of the DNA to be regenerated. Eventually a steady state (K) is reached which largely depends on the wavelength of the incident light (λ):

$$K_\lambda = \frac{\Phi(9)_\lambda \times \varepsilon(\text{Pyr}..\text{Pyr})_\lambda}{\Phi(10)_\lambda \times \varepsilon(\text{Pyr}<>\text{Pyr})_\lambda} \qquad (11)$$

The quantum yield of the forward reaction is of the order of Φ (9) ~ 0.02. The efficiency of the reverse reaction is much higher, Φ (10) ~ 0.8. Hence K_λ decreases with decreasing wavelength. Table 1 shows the ratio of T<>T over T in *E. coli* DNA at different wavelenghts. The ratio of the various pyrimidine dimers depends mainly on the (A-T)/(G-C) content of the microorganism investigated. An example is shown in Table 2.

In addition to the cyclobutane-type dimers, there are other adducts which have a pyrimidine-pyrimidone structure (e.g. Thy(6-4)Pyo **4**, from cytosine + thymine):

4

5

6

Such photoproducts are not photoreversible (note that one molecule of ammonia has been eliminated in the course of the formation of **4**). According to the most recent review on this subject[14], some open questions remain about the structures of these kinds of adducts. In dry spores, a further adduct **5** is observed. It has not yet been observed in other biological systems. The well-known photohydrates (e.g. **6**), although formed with high yields in the monomeric pyrimidine derivatives and single-stranded polynucleotides[16] (and possibly in aqueous DNA[15]), do not appear to be formed *in vivo* in any significant quantities. As with the spore photoproduct **5**, special requirements of DNA structure appear to exist which govern the formation of such photoproducts. In cells with low susceptibility, some Pyr<>Pyr dimer formation is also found at wavelengths where DNA scarcely absorbs (isolated DNA does not give this reaction). It is believed that this must be due to some kind of sensitized reactions. The sensitizer that causes this reaction in cells has not yet been identified.

At present there is great interest in the potential of photosensitized cell killing using specific drugs (sensitizers) to cure cancer. These sensitizers react using mechanisms other than

Table 1: Wavelength dependence of the % T< >T/T ratio in
E. coli DNA in the steady state (cf. Ref. 15).

Wavelength (nm)	% T< >T/T
280	20.0
254	6.5
235	1.7

Table 2: Distribution of Pyr< >Pyr in DNA molecules of different base compositions (A-T/G-C) at a fluence of 200 J m², λ = 265 nm, (cf. Ref. 5).

DNA	$\frac{A-T}{G-C}$	% of Total		
		C < > C	C < > T	T < > T
H. influencae	1.63	5	24	71
E. coli	1.00	7	34	59
M. luteus	0.43	26	55	19

Pyr< >Pyr or Pyr(4-6)Pyo formation. Photosensitization is not likely to play a role in the disinfection of drinking water in the near future. Therefore we will not discuss this question further.

Another process, DNA-protein cross-linking should, however, be mentioned[7,10-11,17]. The inactivation spectrum of most microrganisms has a pronounced peak at 260 nm which is compatible with the importance of DNA as the major target, but there are some examples such as *M. radiodurans*, a microorganism which very effectively repairs Pyr< >Pyr damage, where the maximum of the inactivation spectrum extends to 280 nm, i.e. the maximum of protein absorption. DNA-protein cross-linking has been shown to depend on the microorganism's cell-cycle state. This implies that a particular DNA-protein configuration is required for this reaction to proceed with reasonable efficiency. The chemical nature of such DNA-protein cross-links (lesions) is much more difficult to elucidate than that of the Pyr< >Pyr dimers, since a single such lesion is sufficient to bind the whole DNA to one large protein. Hence, our knowledge about this type of damage is rather limited. Whether or not some recent model studies (cf. Ref. 13) provide good guides remains to be seen.

6. Inactivation of Microorganisms by UV-Radiation. Photoreaction and Repair Processes

The field of photobiology is too large and complex to be dealt with adequately in a few pages. Nevertheless some points which may be relevant for UV-disinfection of surface water will be raised here.

It appears that the major cause of UV-induced inactivation of microorganisms is the formation of the Pyr< >Pyr dimers. Evidence for this comes from the fact that inactivation of a small genome with long wavelength UV-light (λ ~280 nm) can be photoreversed with shorter wavelength UV-light (λ = 240 nm) (see Ref. 18). Pyr(4-6)Pyo lesions are not photoreversible. This does not, however, mean that their formation contributes to lethality. The photoreversibility experiment indicates that Pyr< >Pyr dimers are only damaging lesions.

It is not enough to induce merely one such lesion in the microorganism. Even in a very simple system, such as the simian virus SV40, as many as about 40 dimers are required to prevent its propagation[19]. Dimers are found more or less randomly throughout the whole genome. The propagation of the virus can be blocked for a variety of reasons. Table 3 compiles the contribution (in percent) of the disruption of individual genetic functions to the overall lethal effect.

A virus has no repair system of its own. Therefore the inactivation of viruses appears linear when the log(survival) is plotted versus the fluence (cf. Fig. 8). For cells, this only holds if they are repair deficient. Otherwise an inactivation curve with a more-or-less pronounced shoulder in the low-fluence region is observed (Fig. 9). This has, of course, a bearing on UV-disinfection. Due to the shoulder, the fluence required to reduce survival by the first order of magnitude will be much larger than that required for the next. It is known that the induction of UV-damage triggers an SOS response which, in bacteria, causes increased repair activity, as well as inhibiting division (for references see e.g. Ref. 20). It can also cause mutations (for reviews see Ref. 21 and 22).

Besides such repair reactions which are independent of light, there is another light-induced repair mediated by the photoreactivating enzyme (PRE) present in many microorganisms. PRE recognizes its substrate, the Pyr< >Pyr dimer (equilibrium 12/13). When bound and activated by a photon (λ ~320-480 nm), it can split the dimer:

$$\text{PRE} + \text{Pyr} < > \text{Pyr} \; \underset{(13)}{\overset{(12)}{\rightleftharpoons}} \; \text{PRE} \, .. \, \text{Pyr} < > \text{Pyr} \xrightarrow[(14)]{h\nu} \; \text{PRE} + 2\,\text{Pyr} \; . \qquad (14)$$

As much as 50 - 90 % of the effect of UV-radiation can be reversed by the action of PRE plus visible light. In *E. coli*, one bacterium contains about 20 - 110 PRE molecules depending on the strain (cf. Ref. 23). This low concentration of PRE molecules makes it unlikely that under the conditions of UV-disinfection, i.e. the residence time within the reactor is only a few seconds, considerable photoreactivation will be caused by the light > 320 nm emitted by the disinfecting lamp. This is true even with the high-pressure Hg-arc which emits a major portion of its light in this wavelength region.

Table 4 compiles the fluences required to inactivate different kinds of microorganisms. There is considerable spread of values. This is partly due to the different repair efficiencies, but also to the protection of the DNA by the UV-absorbing material present in the outer part of the microorganisms which prevents much UV-radiation reaching its target, namely DNA. This latter effect is pronounced in the strongly coloured species such as *Aspergillus niger*. The wide range of fluences ensuring sufficient disinfection will have to be taken into account when the hygienic safety of this very promising technique is assessed in detail.

Table 3: UV-irradiation (λ = 254 nm) of the virus SV40. Lethal effect of DNA damage attributable to the disruption of individual genetic functions[19].

Function	Proportion of total lethality (%)
DNA replication	45
RNA synthesis from viral early gene	30
Regulation of early gene transcription	15
Termination of DNA replication: Possible decatenation of interlocked circles	15

Table 4: 90 % reduction level by UV (λ= 254 nm) light (UV-lamp Manufacturers' data, courtesy Heraeus, Hanau).

Microorganism	$J\ m^{-2}$	Microorganism	Colour	$J\ m^{-2}$
Bacteria				
Bacillus anthracis	45	*Staphylococcus albus*		18
B. dysenteriae	22	*Staphylococcus aureus*		26
B. megatherium sp. (veg.)	13	*Streptococcus haemolyticus*		22
B. megatherium sp. (spores)	27	*Streptococcus lactis*		62
B. paratyphosus	32	*Streptococcus viridans*		20
B. subtilis	58			
B. subtilis spores	116	Yeasts		
Corynebacterium diphteriae	34	*Saccharomyces ellipsoides*		60
Eberthella typhosa	21	*Saccharomyces spec.*		80
Escherichia coli	30	*Saccharomyces cerevisiae*		60
Micrococcus candidus	60			
Micrococcus sphaeroides	100			
Neisseria catarrhalis	44	Molds		
Phytomonas tumefaciens	44	*Penicillium roqueforti*	green	130
Proteus vulgaris	30	*Penicillium expansum*	olive	130
Pseudomonas aeruginosa	55	*Penicillium digitatum*	olive	440
Pseudomonas fluorescens	35	*Aspergillus glaucus*	blue-green	440
Salmonella enteritidis	45	*Aspergillus flavus*	yellow-green	600
S. typhimurium	80	*Aspergillus niger*	black	1320
Sarcina lutea	197	*Rhizopus nigricans*	black	1110
Serratia marcescens	24	*Mucor racemosus A*	light grey	170
Shigella paradysenteriae	17	*Mucor racemosus B*	light grey	170
Spirillum rubrum	44	*Oospora lactis*	colourless	50

7. Photochemical Reactions of other Organic and Inorganic Material Present in Surface Water

As mentioned previously, the oxidants chlorine, chlorine dioxide and ozone are capable of degrading most organic material, although there is large variation in the reaction rate constants of these oxidants according to the various substrates that are present in surface water (for ozone see the chapter by J. Hoigné). UV-degradation of high molecular weight material, which is often not easily biodegradable, can be an obstacle in water disinfection because low molecular weight products can then serve as nutrients for the remaining microorganisms and foster their regeneration.

For degradation by UV-light, the first photochemical law always holds: only light which is absorbed can cause photochemical reaction. For this reason, any water treatment prior to UV-irradiation which removes UV-absorbing material will reduce unwanted photochemical side reactions. As a result, the ratio of disinfection to side reactions will be increased.

We can distinguish two classes of photochemical reactions, i.e. direct and sensitized photolyses.

(i) *Direct photolysis*. The effectiveness of direct photolysis depends on the product of the concentration of the compound considered, its extinction coefficient at a particular wavelength, and the quantum yields of the various processes at this wavelength. In a very recent report[4] dealing with the UV-treatment of water, it was stated that the basic process involved in water disinfection and side product formation is the photolytic splitting of molecular oxygen into two oxygen atoms which are then believed to yield OH radicals. These are considered to be active species. Based on gas phase data[24], the 254 nm line of the low-pressure Hg-arc is not absorbed by molecular oxygen (ε (254 nm) $\leq 2 \cdot 10^{-4}$ dm^3 mole^{-1} cm^{-1}). The application of high-pressure Hg-arc lamps is no remedy either. Even at 210 nm, the absorbance by O_2 is very small ((ε(210 nm) $= 7 \cdot 10^{-3}$ dm^3 mole^{-1} cm^{-1}). Bearing in mind that the oxygen concentration in air-saturated water is about $2.5 \cdot 10^{-4}$ mole dm^{-3}, the very small value of the product ε (210 nm) $\cdot [O_2] = 2 \cdot 10^{-6}$ cm^{-1} means that light of this wavelength would penetrate to a depth of 230 m before being attenuated by 10%, leaving 90% of the initial intensity unabsorbed.

(ii) *Sensitized photolysis*. In sensitized photolysis, light is absorbed by a molecule (sensitizer) which is itself not photochemically degraded (photo-catalyst), but transfers a part of the light energy it has absorbed to another molecule. There are too many different photosensitization mechanisms for them to be dealt with in detail in this paper, but some aspects should be mentioned. When the sensitizer acts by energy transfer to a substrate molecule, the contribution of such a reaction does not merely depend on the concentration and extinction coefficient of the sensitizer (see direct photolysis), but also on the concentration of the substrate. Because of the usually short lifetime of the excited states of the sensitizers, energy transfer can only be efficient if the substrate concentration is high. In competition, the excited states of sensitizers are usually readily quenched by molecular oxygen. In general, surface waters are reasonably well oxygenated and this process is expected to play an important role. Among the products of this quenching by molecular oxygen is oxygen in its singlet state. Singlet oxygen is a highly reactive species towards some (but not all) organic substrates, but in water it has only a very short lifetime (~ 3 µs for return to its triplet ground state). Again, this reaction route to side products can only be effective at high substrate concentrations. Through the pretreatment of surface water, the concentration of organic matter is reduced. The remaining organic matter is usually diluted (typically ≤ 3 mg dm^{-3} organic carbon), which means that most (> 99 %) of the singlet oxygen, if formed, returns to its ground state rather than reacting with the organic matter.

Our present knowledge of the photochemical conversion of the organic and inorganic materials present in surface water (photochemical side reactions during disinfection) is still limited. It is expected that the research project "Investigations on the hygienic safety of drinking water disinfection by UV-radiation", sponsored by the German Bundesministerium für Forschung und Technologie, in which our research group is participating, will provide some of the missing information.

One question that we are working on is the photochemical conversion of nitrate (NO_3^-) into nitrite (NO_2^-) ions. Although the mechanisms of nitrite formation are not yet fully understood, it is well-known that this reaction takes place[25-29]. The acceptable limits of nitrate ions in potable water are 50 mg dm^{-3}, while those of nitrite ions are only 0.1 mg dm^{-3}. Thus, if with UV-disinfection only 0.2% of the nitrate allowed to be present is converted into nitrite, the maximum permissible level of nitrite ions is reached. This means that UV-disinfection would not be an acceptable technique. Because of the importance of this question, I would like to discuss some aspects of the photolysis of nitrate and also include some of our own preliminary data on this subject.

Fig. 10 shows the absorption spectrum of the nitrate ion. It can be seen that it has a maximum at 312 nm with a very low extinction coefficient (ε (max) = 7 dm^3 mole^{-1} cm^{-1}). The extinction coefficient is higher below 250 nm, the next maximum being at 185 nm.

Fig. 6 shows that at 254 nm, the main wavelength of the low-pressure Hg-arc, there is merely the onset of the strong far-UV absorption band ε (254 nm) being only 4 dm^3 mole^{-1} cm^{-1}. This means that the light emitted by the low-pressure Hg-arc is not absorbed effectively by the nitrate ions. As mentioned above, strong absorption is one of the requirements for an efficient photochemical reaction. The photolysis of the nitrate ion is surprisingly complex (see

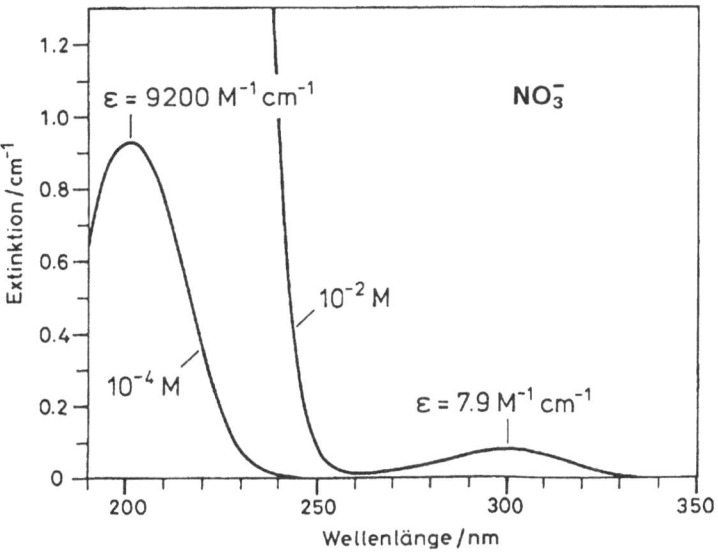

Fig.10:   UV-absorption spectrum of the nitrate ion

previously), but it is clear that its photolysis in the absence of any organic substrate yields nitrite with a quantum yield of only $\Phi(NO_2^-) = 1.5 \cdot 10^{-2}$ mole einstein^{-1}. Due to the complexity of the system, experimental details have to be specified. We have measured the above value of the quantum yield at pH 5.7 ($\pm$ 0.1) and at a nitrate concentration of 10^{-3} mole dm^{-3} (62 mg NO_3^- dm^{-3}) in air-saturated solution. The same number is obtained when the higher values found at high (molar) nitrate concentrations[29] are extrapolated to low nitrate concentrations (these are more relevant for our purposes). The presence of organic matter can have considerable influence on $\Phi(NO_2^-)$. For example, in the presence of 10^{-2} mole dm^{-3} methanol, $\Phi(NO_2^-)$ increases to a value of $5 \cdot 10^{-2}$. These are preliminary experiments, but they indicate the wide spread of $\Phi(NO_2^-)$ values one might obtain upon UV-light treatment of natural water.

One might also wonder if further nitrite is formed by the action of reducing radicals (formed in other photochemical reactions not primarily involving nitrate ions), such as reducing carbohydrate radicals or the superoxide radical anion (O_2^-) which is prevalent whenever free radicals are generated in oxygen-containing aqueous solutions[27]. Such radicals can be readily produced with the help of ionizing radiation[30]. Using this technique we were able to show that such radicals do not convert nitrate to nitrite (see also Ref. 31). Taking now the higher value of $\Phi(NO_2^-) = 5 \cdot 10^{-2}$ mole einstein[31] (in the presence of methanol), we may estimate the amount of nitrite formed upon the action of a fluence that brings the microbial count down from its initial value by four orders of magnitude (for *E. coli* this would be $H = 120$ J m^{-2}; cf. Table 4). In this case, 0.0012% of the nitrate ions have been converted into nitrite ions, i.e. when nitrate is present at a concentration of 50 mg dm^{-3}, nitrite ions are produced at a concentration of $4.5 \cdot 10^{-4}$ mg dm^{-3} (0.45 μg dm^{-3}), which is only 0.5% of the limit set for nitrite in drinking water. In fact, this value is much below the present detection limit of nitrite ions given at 15 μg dm^{-3} (as specified in DIN 38 405, Teil 10). The rough estimates show that nitrite formation from nitrate should not be an obstacle to the use of UV-disinfection in surface water treatment.

At this point, we should examine the emission properties of the two Hg-arcs again. From Figs. 6 and 7, it is evident that the high-pressure Hg-arc emits considerably below 254 nm, while the low-pressure Hg-arc does not. Many organic and inorganic compounds have considerably higher extinction coefficients here than at 254 nm, and many only start to absorb in this wavelength region. This holds, for example, for the nitrate ion (Fig. 10), but also for polymeric carbohydrates (polymers originating from algae, related to alginic acid) found in surface water. For the latter, the chromophores are mainly carboxyl groups, with possibly some contributions from N-acetyl and protein-derived chromophores. Unsubstituted ring-closed carbohydrates do not absorb in the 250 - 230 nm region and the observed photochemistry[32] is due only to a small percentage of open-chain form (C=O function as chromophore) present in the mutarotational equilibrium. In addition, the quantum yield of a given process may go up, in some instances, with decreasing excitation wavelength (e.g. Ref. 33). For this reason, it is thought that high-pressure Hg-arcs produce more side products (at given levels of disinfection) than low-pressure Hg-arcs. Whether this difference matters at all remains to be seen. Some experiments along these lines are to be carried out by our research group. Further side reactions could result from the photochemistry of transition metal complexes or photochemical sensitization of the organic matter by transition metal hydroxides. It is not yet known whether such reactions play a role, but if they do, they will again predominate when the high-pressure Hg-arc is used for disinfection.

In view of these and other open questions, it is premature to favour one or other of the two currently available light sources for disinfection on the basis of possible side reactions induced by them.

8. Will UV-Disinfection be Able to Compete with other Methods ?

As mentioned earlier, there are a considerable number of medium-sized UV-disinfection plants already working. These plants mainly process water whose hygienic quality is already high and show only low UV-absorbance. These are optimal conditions. When the UV-transparency and the hygienic quality of the feed water are low, equipment and electricity costs multiply. Pretreatment of the feed water can certainly increase UV-transparency, but since UV-disinfection is relatively new, little has been done to date to check its technical feasibility. This also holds for UV-lamps and other equipment. Wider use will drastically reduce the price of such equipment and foster the development of cheaper and longer-lived UV-lamps. It is undisputed that, at the moment, disinfection with chlorine is an order of magnitude cheaper than UV-disinfection. The cost difference may decrease with future developments, though UV-disinfection will never be a "cheap" method. The cost is now estimated at roughly 0.02 DM m^{-3}. This appears high, but it is not much when one considers that, in Germany for instance, the consumer pays on average 1.77 DM m^{-3} and in some cities, such as Kassel, the price can rise to 2.75 DM m^{-3} [34]. It should also be mentioned that, in Baden-Württemberg a charge of 0.10 DM m^{-3} is added to the price of every m^3 of drinking water as "Wasserpfennig" to compensate for the losses the farmers encounter when they reduce the use of artificial and organic fertilizers in water catchment areas [35]. This is meant to ensure the reduction of nitrate ion concentration in the drinking water. Thus it can be anticipated that consumers will accept paying an increase of 0.02 DM (ca. 1%) for UV-disinfection if, in return, they receive water free of chlorine and chlorinated compounds. Chlorine dioxide does not produce chlorinated compounds, but its reduction product, chlorite ClO_2^-, is by no means unproblematic. The use of chlorine dioxide has therefore been limited to 0.3 mg dm^{-3}, and, as mentioned above, this concentration is often insufficient to guarantee adequate disinfection. Generally ozone cannot act as a substitute for these disinfectants and some water tastes " off " when treated with ozone (see previously).

One of the arguments in favour of oxidizing disinfectants has always been that they can be used in large enough quantities to be still present in low concentrations at the consumer's end and hence guarantee stable disinfection throughout the whole network. Not all agree with this. Large cities, such as Munich, which have high quality water supplies, refrain from chlorination of their pipeline systems. Even the city of Amsterdam, whose water supply stems from processed Rhine water, no longer considers the standard level of chlorine to be a hygienically beneficial measure. Hence, it may well be that UV-disinfection will soon become one of the most powerful tools for producing high-quality drinking water at reasonable cost.

Acknowledgement. Some of the work mentioned in the text was sponsored by the German Bundesministerium für Forschung und Technologie, Project 326-4002-02-WT 8720 3.

References

1 Henri, V., Helbronner, A. and de Recklinghausen, M., Nouvelles recherches sur la stérilisation de grandes quantités d'eau par les rayons ultraviolets. Compt. Rend. Acad. Sci. **151** (1910) 677.

2 Perkin, F.M., Mercury vapour lamps and action of ultra violet rays. Trans. Faraday Soc. **6** (1910) 199.

3 Henry, V., Helbronner, A. and de Recklinghausen, M., Stérilisation de grandes quantités d'eau par les rayons ultraviolets. Compt. Rend. Acad. Sci. **150** (1910) 932.

4 Thiemann, W., Bericht über die Tagung "Ozon und Ultraviolett-Wasserbehandlung". GWF Wasser Abwasser **128** (1987) 441.

5 von Sonntag, C., Disinfection by free radicals and UV-radiation. Wat. Supply **4** (1986) 11.

6 S.Y. Wang, ed., *Photochemistry and Photobiology of Nucleic Acids*, Vol. 1 and 2 (Academic Press, New York) 1976.

7 Sperling, J. and Havron, A., "Specificity of photochemical cross-linking in protein-nucleic acid complexes" in *Excited states in Organic Chemistry and Biochemistry*, by B. Pullman, N. Goldblum eds. (Reidel, Boston) 1977, p. 79.

8 Löber, G. and Kittler, L., Selected topics in photochemistry of nucleic acids. Recent results and perspectives. Photochem. Photobiol. **25** (1977) 215.

9 Rahn, R.O., Nondimer damage in deoxyribonucleic acids caused by ultraviolet radiation. Photochem. Photobiol. Rev. **4** (1979) 267.

10 Matsuura, T., Saito, I., Ito, S., Sugiyama, H. and Shinmura, T., Organic chemical approach to photo-cross-links of nucleic acids to proteins. Pure Appl. Chem. **52** (1980) 2705.

11 Shetlar, M.D., Cross-linking of proteins to nucleic acids by ultraviolet light, Photochem. Photobiol. Rev. **5** (1980) 105.

12 Saito, I., Sugiyama, H. and Matsuura, T., Photochemical reactions of nucleic acids and their constituents of photobiological relevance. Photochem. Photobiol. **38** (1983) 735.

13 Cadet, J., Voituriez, L., Grand, A., Hruska, F.E., Vigny, P. and L.-S Kan, Recent aspects of the photochemistry of nucleic acids and related compounds. Biochimie **67** (1985) 277.

14 Cadet, J., Berger, M., Decarroz, D., Wagner, J.R., van Lier, J.E., Ginot, Y.M. and Vigny, P., Photosensitized reactions of nucleic acids. Biochimie **68** (1986) 813.

15 Patrick, M.H. and Rahn, R.O., "Photochemistry of DNA and polynucleotides: photoproducts" in *Photochemistry and Photobiology of Nucleic Acids*, Vol.2, by S.Y.Wang, ed. (Academic Press, New York) 1976, p. 35.

16 Fisher, G.J. and Johns, H.E., "Pyrimidine photohydrates" in *Photochemistry and Photobiology of Nucleic Acids*, Vol. 1., by S.Y. Wang, ed. (Academic Press, New York) 1976, p. 169.

17 Smith, K.C., "The radiation-induced addition of proteins and other molecules to nucleic acids" in *Photochemistry and Photobiology of Nucleic Acids*, Vol. 2, by S.Y. Wang, ed. (Academic Press, New York) 1976, p. 187.

18 Jagger, J. "Ultraviolet inactivation of biological systems" in *Photochemistry and Photobiology of Nucleic Acids*, Vol. 2, by S.Y. Wang, ed. (Academic Press, New York) 1976, p. 147.

19 Brown, T.C. and Cerutti, P.A., Ultraviolet radiation inactivates SV40 by disrupting at least four genetic functions. EMBO J. **5** (1986) 197.

20 Caldeira de Araujo, A. and Favre, A, "Near ultraviolet DNA damage induces the SOS responses" in *Escherichia coli*, EMBO J. **5** (1986) 175.

21 Doudney, C.O., "Mutation in ultraviolet light-damaged microorganisms" in *Photochemistry and Photobiology of Nucleic Acids*, Vol.2, by S.Y. Wang ed. (Academic Press, New York) 1976, p. 309.

22 Hutchinson, F., Yearly review. A review of some topics concerning mutagenesis by ultraviolet light. Photochem. Photobiol. **45** (1987) 897.

23 Harm, H., "Repair of UV-irradiated biological systems: photoreactivation" in *Photochemistry and Photobiology of Nucleic Acids*, Vol.2, by S.Y. Wang, ed. (Academic Press, New York) 1976, p. 219.

24 Calvert, J.G. and Pitts, Jr., N., *Photochemistry* (Wiley and Sons, New York) 1966, p. 207.

25 Daniels, M., Meyers, R.V. and Belardo, E.V., Photochemistry of the aqueous nitrate system. I. Excitation in the 300-mµ band. J. Phys. Chem. **72** (1968) 389.

26 Shuali, U., Ottolenghi, M., Rabani, J. and Yelin, Z., On the photochemistry of aqueous nitrate solutions excited

in the 195-nm band. J. Phys. Chem. **73** (1969) 3445.

27 Bayliss, N.S. and Bucat, R.B., The photolysis of aqueous nitrate solutions. Aust. J. Chem. **28** (1975) 1865.

28 Wagner, I., Strehlow, H. and Busse, G., Flash photolysis of nitrate ions in aqueous solution. Z. Physik Chem. N.F. **123** (1980) 1.

29 Dubovitskii, A.V., Leksina, L.N. and Manelis, G.B., Photolysis of nitrate ions in aqueous solutions. High Energy Chem. **15** (1981) 265.

30 von Sonntag, C., *The Chemical Basis of Radiation Biology* (Taylor and Francis, London) 1987.

31 Elliot, A.J. and Sopchyshyn, F.C., The radiolysis at room temperature and 118 °C of aqueous solutions containing sodium nitrate and either sodium formate or 2-propanol, Can. J. Chem. **61** (1983) 1578.

32 Triantaphylides, C., Schuchmann, H.-P. and von Sonntag, C., Photolysis of D-fructose in aqueous solution. Carbohydr. Res. **100** (1982) 131.

33 Campbell, J.M., von Sonntag, C. and Schule-Frohlinde, D., Photolysis of 5-bromouracil and some related compounds in solution. Z. Naturforsch. **29b** (1974) 750.

34 Frankfurter Allgemeine Zeitung, August 22, 1987, p. 12.

35 Frankfurter Allgemeine Zeitung, August 25, 1987, p. 5.

Discussion

Chairman: Dr. E. Merz
Brown Boveri, Zürich, Switzerland

W.A. McRae   •   You mentioned that, if the DNA absorbs UV light producing about 40 dimers, it can effectively not be repaired. So, if we assume we've got a sterilization lamp which is well operated and we make sure that all bacteria get their 40 dimers lesions, then we are alright. Let's suppose we have a poorly operated system where the bacteria are not killed but there are changes in the DNA. Do we risk mutations in the micro-organisms? We know, for example, UV irradiation may lead to mutations in human skin.

C. von Sonntag   •   Some mutations. This is inevitable with UV light and with any sort of disinfectant that you might be using. There is always a risk of forming mutations using UV light or any sort of disinfectant.

W.A. McRae   •   Do these mutations then have to be studied to see if they've made the situation worse rather than better?

C. von Sonntag   •   I don't think so. Most of the mutations will produce changes in some sort of gene expressions. As far as I know, the formation of any more dangerous mutations has never been reported.

M. Campagna   •   I have a question which is somewhat related to what has been just mentioned. I was wondering how universal your mechanism is related to the fluence. I didn't see units on your scale (Figs. 8 and 9) so I was wondering whether you would expect that repair mechanism could be scaled somehow. If you know what is happening, then you just need to adjust the power of your lamp.

C. von Sonntag   •   The reason why I didn't put any scales is that, depending on the kind of microorganism, the dose required depends on various factors. One factor is whether or not the bacterium or the mould is coloured. Another factor is that the different strains of bacteria have different repair capabilities. One reason for this is that some bacteria have more than one genome. So if you damage one genome, they still have the other available and can use it as a template. The quantitative extent of the shoulder will depend very strongly on the kind of bacterium being treated. Some bacteria have a very pronounced shoulder, i.e. an efficient repair system. Some strains of the same bacteria might not even produce a shoulder at all because they lack some repair enzymes. That was the reason why I kept it general and qualitative. So what we have to do is to find out the kind of bacteria present, make a very careful microbiological study and then adjust the fluence according to the needs. Adjustments may also be necessary during the year if surface water is treated, because different bacteria may predominate at different seasons. What, of course, can be done is to adjust the fluence to the highest value necessary allowing some overkill during the other periods.

E.A. van Naerssen   •   UV is very effective against bacteria but it seems not to be that effective against higher organisms like rotifera, latisfera or even worms like nematura probably because the cell walls - in this case the skin - of the organisms are not transparent enough to UV light. Can you see any developments in future that UV combined with something else. e.g. UV in the presence of ozone, could attack these higher organisms? This would be a technique of great value in water treatment.

C. von Sonntag • I'm not a specialist in the field of hygiene, but I think you are right about the absorption of UV light in these higher microorganisms. The organism will not be killed just by destroying the DNA of one cell because there are many cells. To kill the whole organism will take a much higher fluence. I don't know much about the problem of the disinfection of those microorganisms that you are talking about. As I mentioned before, I'm a chemist and a newcomer to the field.

W. Lorch • My question relates specifically to the influence of UV treatment on potable water as far as further upgrading is concerned. One of the greatest concerns for upgrading water for pharmaceutical and medical purposes is the presence of pyrogens. Now you explained clearly that you achieve - I'm using your own terminology - a total kill and inactivation. What happens to the breakdown product clearly shown on your slides? Do they remain in the solution? Because if so, pyrogens, temperature raising identities, will remain. If they did not, it would have a tremendous influence on the further upgrading of water for injections.

C. von Sonntag • In order to reduce the pyrogens, we would have to degrade the pyrogens photolytically. The pyrogens however might not absorb light adequately. As far as I know, pyrogens are proteins. Now the first problem is that proteins do not absorb the UV light as strongly as DNA. The second problem might be that their photochemical destruction is not as effective and small changes caused by photochemical alterations might not reduce the pyrogen activity. You might have to go to much much higher doses to effectively do this.

W. Lorch • To be quite productive may I ask if any pyrogen tests have been carried out on water treated by UV irradiation? I think it'd be worthwhile and productive to do so.

C. von Sonntag • I think your old friend, Prof. G.O. Schenk, is trying to induce such a study.

P. Francis • I could talk a lot about pyrogens, UV and ozone but I'll keep that to one side for the minute. I would like to comment on the shape of the inactivation curve of the bacteria. We find that, with heavy loadings in a batch type reactor with typical residence times of say 10 minutes, you get a rapid kill and then a shoulder. What do you think might be the secondary shoulder?

C. von Sonntag • This behaviour has been observed for all disinfectants including chlorine. In a first rapid reaction, most of the bacteria are inactivated. What remains then are the very repair efficient bacteria. That is one of the explanations. There is also the possibility that some bacteria are protected from the action of UV light by e.g. mineral particles present in the water. They hide in the shadows.

High Purity Water for Semiconductor Manufacturing

E. Krapf and R. Preisser
IBM, Sindelfingen, West-Germany

1. Introduction

Ultrapure water is an essential chemical and material for the production of semiconductor devices. Impurities in the process water such as bacteria, organics, ionic contaminations and particles are a potential cause of reduced device yield and reliability failures.

In order to reach the standards set today and in the future, it is essential to optimize the complete system of water purification including the plant's hardware as well as its mode of operation.

2. The Importance of Ultrapure Water in Semiconductor Processing

All wet processing steps in microelectronic manufacturing using inorganic or organic solvents need a post-rinse with deionized water in order to stop surface reactions and to neutralize the surface. Different wafer surface structures and materials, such as silicon dioxide, doped or undoped silicon, nitride or poly-silicon films interact specifically with the treatment solutions and the rinsing water.

Wafers with small geometries and steep slopes, as they typically occur in the etching structures of microelectronic devices, require other rinsing conditions than non-structured wafers.

The total ultrapure water consumption per wafer and processing step is estimated to be in the order of 5 to 10 liters. This adds up to 1000 to 5000 l per wafer for a typical process sequence in microelectronic manufacturing.

A variety of techniques is used to bring the wafers into contact with the rinsing water, depending on the surface conditions of the wafers: tank-dip-cleans, spray cleans, high pressure spray cleans, hot vs. warm cleans, etc.

The impurities present in the semiconductor rinsing water are generally classified according to the groups given below:

- Microorganisms
- Sub-micron Particles
- Silica
- Ionic Contaminants
- Organic Materials
- Dissolved Oxygen

The water impurities can have different effects on the semiconductor devices:

- Degradation of the breakdown voltage strength of dielectric films (MOS structures), indicating that the films contain impurities from the processing. The extent of the damage can be quantified by a "film integrity test" (see Fig. 1). The uniformity of the breakdown voltage is a function of the film contamination. The histogram of Fig. 1 shows a narrow distribution of the breakdown voltage, i.e. the film is uniformly clean. The spread of low breakdown voltages in Fig. 2 indicates localized contamination of the device.

- Surface or junction leakage is caused by ionic impurities. The leakage depends on the diffusion coefficient of the contaminating ions and the contaminated material.

- Cross defects caused by particulates. In conjunction with subsequent process steps this may lead to missing patterns or bridging between islands.

- Adsorbed contaminants, localized or on large areas, can lead to problems with respect to grain boundaries or adhesion.

- Oxidizing impurities or halogens may cause corrosion effects.

The degree of purity of the water in the distribution system has always to be adapted to the most critical process using the water. The reason for this is that additional purification close to the point of use of the critical manufacturing process is only to a very limited extent, feasible for most of the contaminants. The quality requirements of the most critical process step hence set the standards for the whole water supply system.

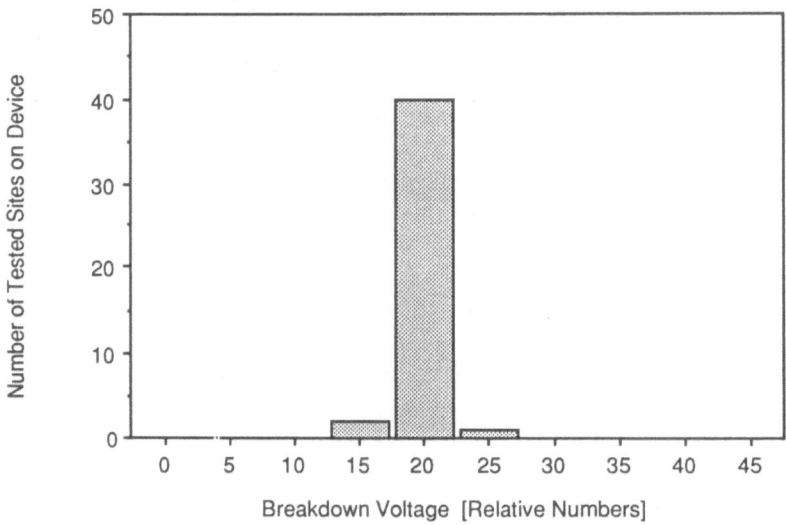

Fig. 1:   Contamination control by film integrity test. The breakdown voltage of a MOS film is measured at a number of different sites on the device. Contaminations lead to a reduced breakdown voltage. Example of a satisfactory film, characterized by a narrow distribution of breakdown voltages.

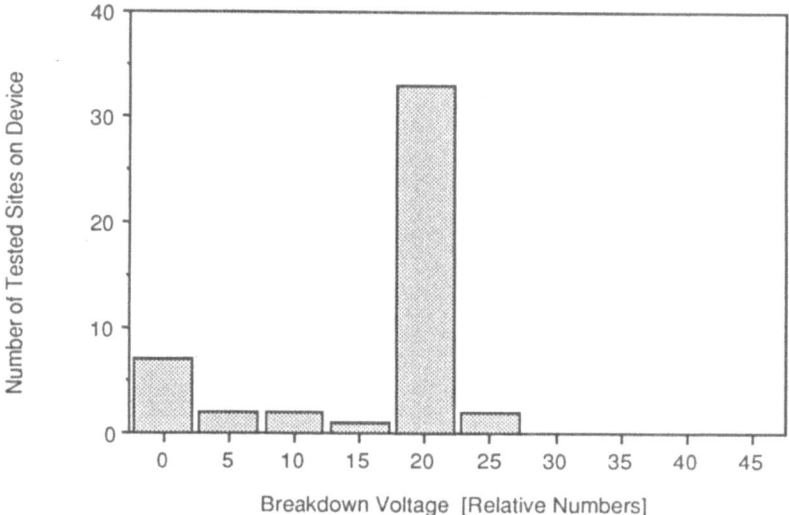

Breakdown Voltage [Relative Numbers]

Fig. 2:   Contamination control by film integrity test. The breakdown voltage of a MOS film is measured at a number of different sites on the device. Contaminations lead to a reduced breakdown voltage. Example of a contaminated device: the distribution of breakdown voltages shows a scatter towards lower values.

3. Technology of Ultrapure Water Preparation

The preparation of ultrapure water from raw water follows a sequence of more or less established process steps. A typical sequence is shown in Fig. 3. Fig. 4 gives a more detailed flow-sheet of the final purification stages of the system (i.e. the treatment processes below the dotted line in Fig. 3), as used at the IBM plant in Sindelfingen.

4. Quality Assurance

The most sensitive monitor for contaminations is the semiconductor device itself. Correlations between contaminant concentrations and device yield exist. Sensitive monitoring of impurity concentrations is, however, required to protect the costly manufacturing line from producing large quantities of default devices in case of water contamination. A number of analytical techniques have been applied to monitor the quality of the water in the polishing loop and at the point of use (Fig. 4).

Inorganics: Metallic impurities in general, or specific ones, such as Na, Fe, etc., are monitored by inductively coupled plasma spectroscopy (ICP). Typical metal concentrations tolerated for current semiconductor applications are in the sub-ppb range. The limiting values for the contaminants are given in Table 1. If the tolerated levels of metal impurities are exceeded, the production has to be stopped immediately. To remove the contaminants from the network, the system must be purged.

Resistivity measurements using conductivity meters with automatic temperature compensation are used for on-line monitoring of ionic impurities.

Table 1:　Quality of high purity water.

Parameter	Upper Control Limit
Resistivity [MΩ cm]	> 17.5
Bacteria [germs/m*l*]	
- Life Millipore	200
- Total EPI	1000
- Total SEM	1000
- Pyrogene [ng/*l*]	2.5
DOC [ppb]	≤ 10
Ionic Contamination [μg/*l*]	
- Cations (ESA/ICP)	5 (each)
- Anions (IC)	1 (each)
- Na	1
- Si	1
Particles [count/*l*]	
- OLC　(> 0.5 μm)	500
- SEM　(> 0.5 μm)	2000
(> 0.2 μm)	10000

Anion contaminations are monitored by ion chromatography (IC). Excessive levels may lead to corrosion problems and to particle build-up. Modern IC can be used as a semi-on-line analytical method and is sensitive even to low ppb levels.

Microorganisms and organics: The counting of live bacteria still relies on culturing methods (standard plate counts). These methods are, however, inconveniently time consuming if immediate reaction to a microbial contamination is called for. The method is used in addition to epifluorescence, which measures numbers which are roughly proportional to the number of bacteria present. Epifluorescence measurements, however, rely on the uptake of a dye by the bacteria, the amount of which can vary for different bacteria. As can be seen from the histogram in Fig. 5, the results obtained from plate counts and from epifluorescence do not correlate for this reason. If very low levels of organic particles have to be monitored, pyrogen tests can be used. Pyrogen tests are not used throughout the whole plant, but at selected points, such as RO and ultrafiltration units.

DOC (Dissolved Organic Carbon) is monitored using on-line equipment to measure the total amount of organic impurities.

If bacteria and organics exceed the tolerated limits, the system needs to be purged and, additionally, disinfection may become necessary. In the future, in-situ ozonation will be used to control microbial contaminations.

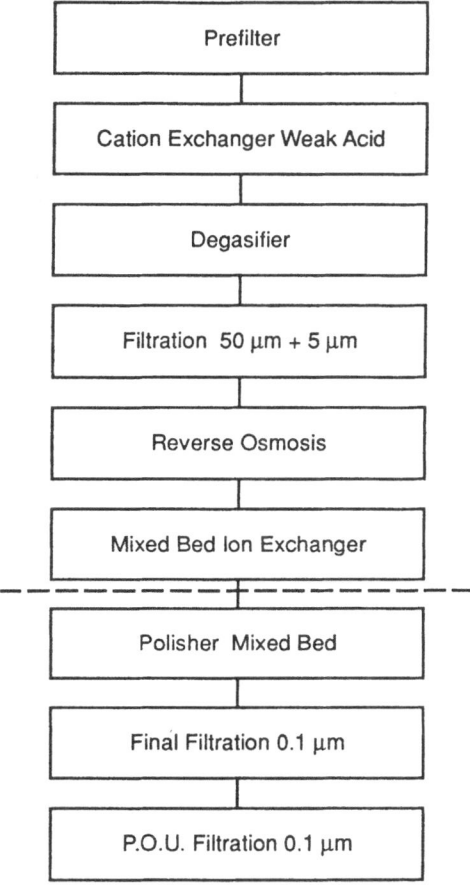

Fig. 3:   Process flow chart of an ultrapure water preparation plant for semiconductor manufacturing.

Micro-particles: The number of micro-particles of organic or inorganic origin is monitored by an optical measurement system using He/Ne - lasers. Particle sizes of 0.3 to 0.5 µm can be detected by current technology. In order to keep the particle counts low, 0.1 µm filters are used at different points in the plant. Particle analysis for future systems aims at the detection of particles in the 0.1 µm range and at on-line operation. Data measured by the optical systems are referred to as "particle equivalent counts"; a standardized calibration method does not exist to date. Scanning electron microscopy has been used so far to support or calibrate the laser interferometric analysis results. SEM is necessarily an off-line technique. It counts the number of particles collected on a 0.1 µm filter after filtration of a given sample volume. Fig. 6 shows a comparison of the results obtained using the different particle counting techniques.

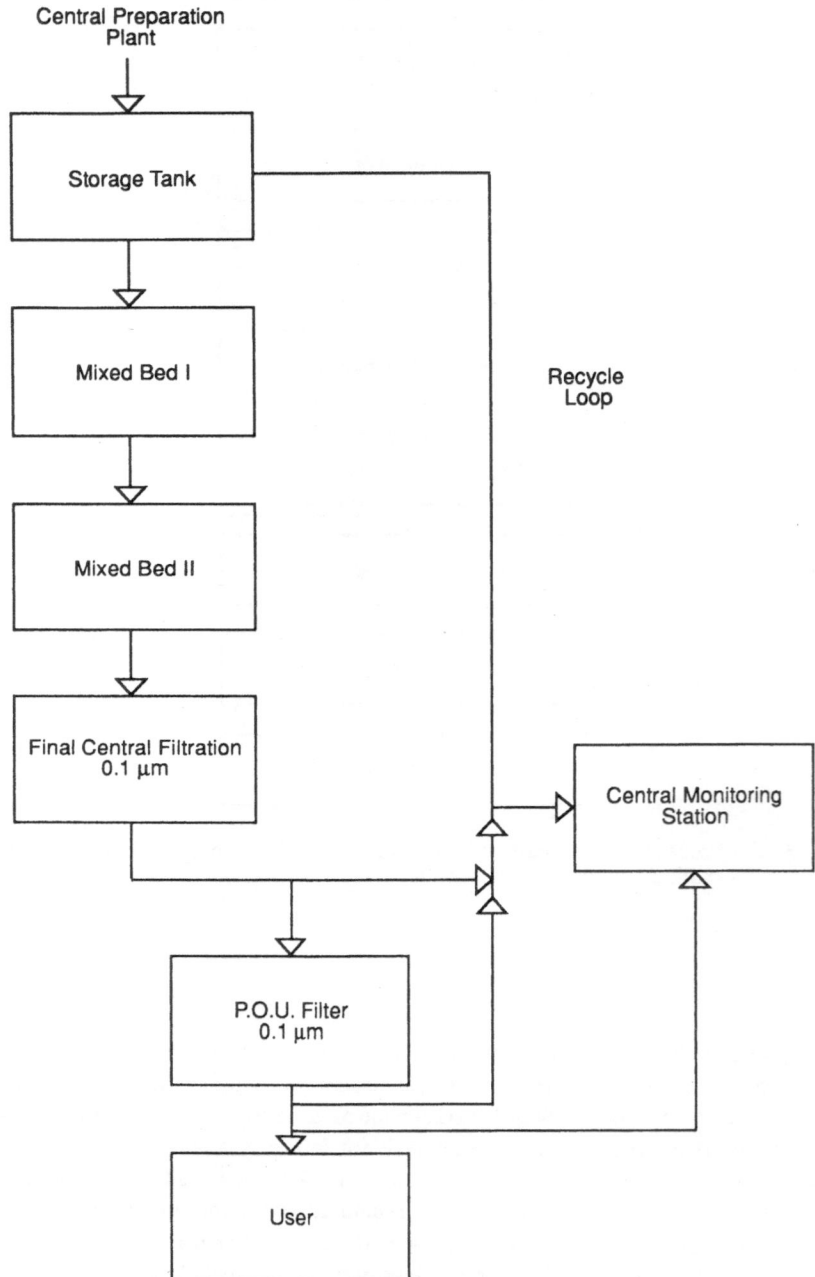

Fig . 4:   Final treatment system of an ultrapure water plant.

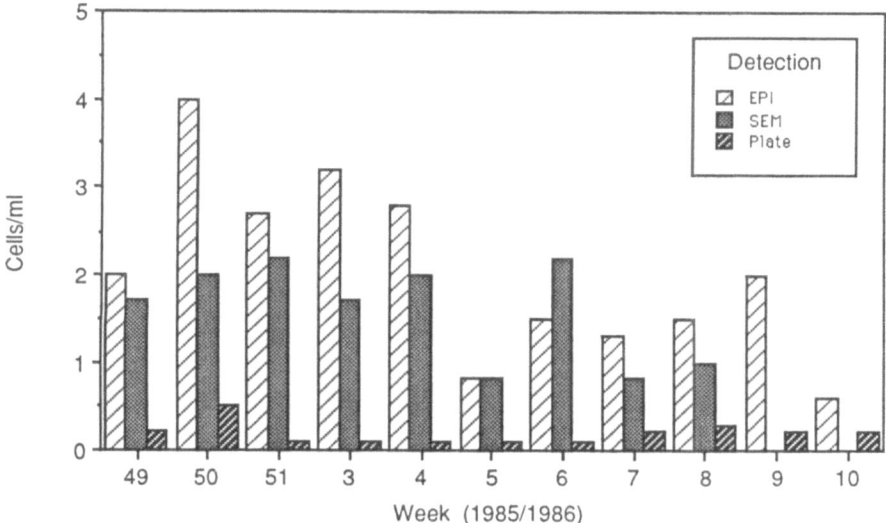

Fig. 5: Comparison of various techniques used to assess microbial contamination in ultrapure water from a typical production plant: Epifluorescence and SEM (scanning electron microscope) both rely on filtering a large volume of the water to be tested (1000 *l*) through a disc filter and analyzing the residue. SEM allows distinction between inorganic and organic material, whereas epifluorescence measures total number of bacteria by a specific dye reaction.

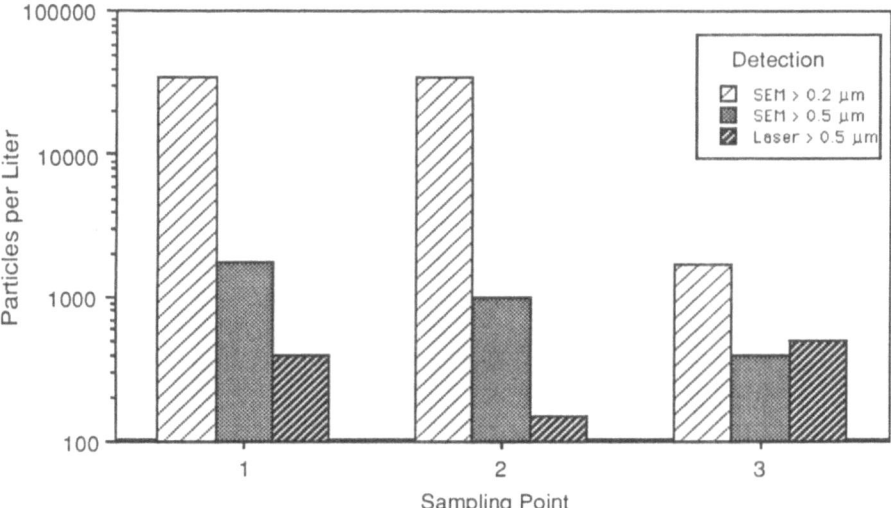

Fig. 6: Comparison of SEM counting using 0.2 and 0.5 µm filters with laser method for measuring the number of particles in samples taken at three different points in the ultrapure water plant.

5. Summary and Outlook

The design and construction of the ultrapure water plant are crucial for the quality of the product water. The important design criteria are: piping size, materials, layout of the plant, flow pattern and pressure. The choice of construction materials is particularly important to ensure that the system is free of major contaminants. PVDF has proved to be superior to PVC with respect to particulates and DOC. Bacterial growth can only be controlled in flowing systems. It is necessary to make sure that all dead-ends are removed from the installation. Filters, placed at different points in the plant, assure a stable particulate level independent of time variations in water demand.

Extensive monitoring of the water quality is required in order to assure a minimum risk of contamination and hence reduced manufacturing yield.

In the future, the trend will be to use the following techniques: in-situ disinfection techniques, advanced filtration using filters with smaller pores, surface activated membranes, and ultrafiltration at the point of use in order to meet the requirements of VLSI and ULSI manufacturing. Improvements in the water flow and the rinsing techniques themselves will further increase the current capabilities.

Discussion

Chairman: E. Merz
Brown Boveri, Zürich, Switzerland

G. Schock • You pointed out the high quantity of water that is used in your production lines. Do you use any water reclamation systems and if no, why and if yes, what is the amount of recycled water? What are the purification steps and to which point of your water system is the water recycled?

R. Preisser • We have recycled water but it is never used for rinsing wafers or other products. For this purpose we always use pure water that went through all the purification steps.

R. Rautenbach • Maybe it would be easier to regenerate ultrapure water from your recycled rinsing water than from ordinary tap water.

R. Preisser • I'm not sure about this. Because the solubility of some of the important traces, like the dopants in the films, are not known to the extent which would be necessary to make sure it is not a danger for the virgin product. A contaminant dissolving out of the semiconductor in the ppt range, may affect the device, but would never be detected by the installed analytical equipment. That's the reason why we try to avoid such specific contaminations. On the other hand, you are right, it's an economical question and we recycle some of the water, but not on the virgin device.

C. Schüler • You have shown us the extraordinary requirements of your plants. Now I would be interested to learn, to what extent you can satisfy these requirements by buying the equipment from outside or getting the engineering done from outside and to what extent you have to do that yourself?

R. Preisser • We define the layout of the equipment ourselves and buy all the components from the outside. We examine new developments and try out, how they can fit our strategy. If there is a better alternative (e.g. ozonation) we will test it and afterwards decide what we are going to do.

C. Schüler • So you ensure your own quality in your factory and you don't turn that over to the suppliers of the plant.

R. Preisser • That's exactly right.

W. McRae • Could you give us some idea of the cost of a m³ of the ultrapure water that you use?

R. Preisser • It's somewhere around 5 to 6 DM per m³.

M. Campagna • What is the total capacity of such installations worldwide?

R. Preisser • IBM is producing memory chips and logic chips in different sites. On these sites we have a corporate strategy and this was reflected in my paper. I'm not aware of the details of plants of other semiconductor manufacturers.

M. Bodmer   •   There is a very high similarity between your monitoring system and the monitoring system used in a power station, especially a nuclear power station where the concentrations to be monitored are exactly like you told us: at the limits of detection. My question is: In condensate polishing using mixed-bed units we have a problem caused by cross contamination of cations and anions during separation prior to regeneration. The trend is now to increase the purity of the polished water by using separated cation and anion exchangers without mixed beds. Did you consider these methods?

R. Preisser   •   No, we did not consider these techniques. Currently we are focussing on our existing system and try to clean this system up as close to the point of use as possible. This is essential because it has been found that among the impurities which are detectable at the point of use, there are no major components traceable back to the early stages of the purification of the water.

M.C. Kavanaugh •   When you have recordings in your quality control that exceed your limits what is your strategy?

R. Preisser   •   We have different strategies. It depends on what impurity in what quantity is causing the alarm. It can mean stopping the production at a special place, if the problem is localized, or, if the problem is not localized, stopping the complete production line. After shut-down we have a few different techniques. The simplest one is to rinse the system and to dump the water.

Application of Electrochemical Ozone Generators in Ultrapure Water Systems

S. Stucki and H. Baumann
Brown Boveri, Baden, Switzerland

1. Introduction

Manufacturing processes in the microelectronic, pharmaceutical and related industries need large quantities of purified water. While classical ion exchange combined with membrane desalination processes, such as reverse osmosis (RO) or electrodialysis (ED), is unproblematic in yielding very low levels of ionic impurities, the efficient elimination of non ionic trace impurities still causes problems in ultrapure water plants. The non ionic impurities are normally classified as TOC (Total Organic Carbon = the sum of dissolved or suspended organic matter) particles and living microorganisms have been found to critically influence the product yield in microelectronics manufacturing. In some pharmaceutical applications (e.g. water for injections, and some biotechnical processes), sterility, i.e. the absence of any microorganisms has to be guaranteed.

To produce water of this quality, additional purification processes have to be included in the recirculating (or polishing) loop of the ultrapure water plant. Point of use filtration using UF or even RO membranes[1,2] is capable of substantially reducing bacteria and particle counts. Whether TOC levels can be reduced via membrane filtration strongly depends on the nature of the organic contamination, its molecular weight and the properties of the membrane module.

The use of ozone as a disinfectant/oxidant in ultrapure water has become an interesting alternative or supplement to UV treatment[3,4]. The advantage of ozone over other chemical oxidants is its intrinsic instability. This leads eventually to the formation of molecular dioxygen, which is not considered to be a contaminant for most applications, but if necessary it can easily be removed. Ozone has been found to be an active disinfectant in concentrations as low as 20 ppb[3,4]. Ozone for use in ultrapure water systems must be free of contaminants other than O_2, i.e. the ozone generator should not produce any side-products. This practically excludes the use of air-fed ozonizers for this purpose, since the formation of trace amounts of NO_x cannot be avoided.

2. Membrane Electrolytic Ozone Generators

A new electrochemical cell has been developed at Brown Boveri for the production of ozone from pure water[4,5,6,7]. The cell consists of a chemically stable perfluorinated proton exchange membrane (Nafion®, Du Pont), contacted by corrosion resistant porous electrodes. Water is pumped through the anode compartment of the cell. A small part of the water is electrochemically split at the specially designed high overvoltage anode. The anodic reaction

Table 1:   Technical performance of Membrel® ozone generators.

Operating Temperature	10...50	°C
Current Density	< 2	A/cm²
Ozone Yield	< 20	wt%
Production Rate	< 12	g/(h dm²)
Cell Voltage	< 5	V
Energy Efficiency	> 60	kWh/kg

product is a mixture of oxygen with up to 20% wt. ozone. This represents an ozone generator with an unusually high concentration of ozone in the product gas. The high ozone concentration enables a fast and efficient transfer of ozone into the water which passes through the cell. The production and contacting of the ozone are effected by one and same piece of equipment, which makes it suitable for operation "in situ".

Table 1 summarizes the technical performance of the Membrel® electrolytic ozone generator.

The combination of the solid electrolyte with a high overvoltage PbO_2/Titanium anode has proved to be compatible with ultrapure water. Analysis of the water after prolonged electrolysis in a recycling mode has shown that the specific corrosion of the cell components is not detectable for Pb and smaller than 10^{-6} g/g ozone for F⁻. Cells have been on constant operation at a current density of 1 Acm⁻² for several years without losses in performance. These results have enabled various applications of Membrel ozonizers in water purification plants.

3. Application of Membrel® Ozonizers for the Disinfection of Ultrapure Water

The growth of microorganisms in ultrapure water systems adversely affects the high quality of the product water. The problem is particularly severe in large distributing systems with hundreds of meters of piping as well as in large buffer tanks. For ultrapure water systems in pharmaceutical and microelectronic production various processes have been applied to control microbial contamination. The most commonly used process is UV irradiation, which has the advantage of being non-chemical in the sense that no chemicals are added. UV disinfection has the disadvantage in that it is only active in the irradiated volume, while microorganisms may still survive in the shade[3]. Both chemical disinfectants in high doses in periodical application and steam sterilization require shut-down of the plant, which may conflict with the manufacturing processes.

Ozone, and in particular ozone generated in situ by Membrel electrochemical cells, has been successfully introduced into ultrapure recycle loops to control bacterial growth[4,6]. Fig. 1 shows a simplified flow sheet of such an ozonation plant. The ozone generator is usually placed in a bypass stream of the main recycle stream of the ultrapure water distribution system. It was found and proved in practice that maintaining a permanent ozone concentration in the water as low as 20 to 100 ppb is sufficient to guarantee sterility in the water[4]. If necessary, trace amounts of ozone in the product water can be destroyed very easily by UV radiation at the point of use. Some 25 ultrapure water plants have been equipped with Membrel electrolytic ozone generators over the past few years.

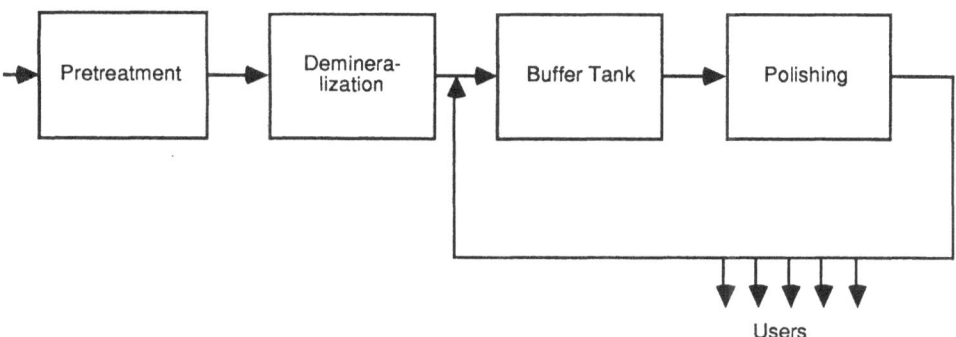

Fig. 1:   Simplified flow-sheet of a water purification plant.

4. Application of Membrel Ozonizers for TOC Elimination from Ultrapure Water

State of the art demineralization processes produce water containing less than 10 ppb of inorganic contaminants (ions and silica). In comparison, organic contaminants are typically present in concentrations one order of magnitude higher and therefore represent the highest mass percentage (90% and more) of the totality of ultrapure water contaminants[8]. The concentrations, measured as ppb TOC, are mainly given by the retention of the reverse osmosis membranes for organics and by organic matter leaching out of water purification equipment (e.g. ion-exchange resins and membrane modules) and other polymer material in contact with the water (e.g. piping).

The tolerated concentration levels of TOC in water for microelectronic manufacturing have come down from 500 ppb in 1979 to 50 ppb to date. Various processes have been developed to reduce TOC concentrations in ultrapure water[9].

RO removes most of the organic contaminants present in raw water, such as most of the humic acids and other large molecules. The retention of RO membranes for small molecules depends strongly on the chemical nature of the molecules and the membrane material. Ultrafiltration and, more recently, RO have been used for point-of-use filtration[1,2].

Oxidative treatment of the water with 185 nm UV has been proposed and shown to be effective[8], although 185 nm UV in large scale applications seems to be too expensive with state-of-the-art lamp technology. The idea behind oxidative treatment for TOC elimination is that, by oxidation in water, all organic molecules are eventually transformed into ionogenic entities like carboxylic acids or even carbon dioxide. Ions are easily removed from the water by the mixed bed polishers, which are standard in any ultrapure recycle loop. The oxidation reaction has to be sufficiently fast for the oxidative treatment to become technically feasible. It is well known from the work of Hoigné and co-workers[10] that the direct oxidation rate of ozone with different organic molecules can vary by several orders of magnitude. On the other hand, oxidation with ozone can also proceed via radical intermediates, mainly OH. This reaction type is known to be non-specific and fast. For the oxidative elimination of TOC in ultrapure water, the radical reaction path seems to be the only practical way since the nature of the organic contaminants is generally not know. The radical oxidation path is known to be promoted among others by high pH (i.e. by OH$^-$-ions), by hydrogen peroxide and ultraviolet (254 nm) radiation. Since pH cannot be used as a parameter in pure water, we studied radical oxidation with the combinations UV/ozone and H_2O_2/ozone as the source of OH radicals.

The feasibility of the process was studied using a laboratory test loop as shown in Fig. 2. The loop consists of an in situ electrochemical ozone generator, a stirred "ozone" reactor with optional H_2O_2 dosage, a photochemical reactor with a 254 nm UV lamp and a mixed bed polisher including a filter for particulates. The instrumentation includes conductivity meters and on-line analyzers for dissolved ozone and for TOC. Experimental details of the test loop are given in Ref.11.

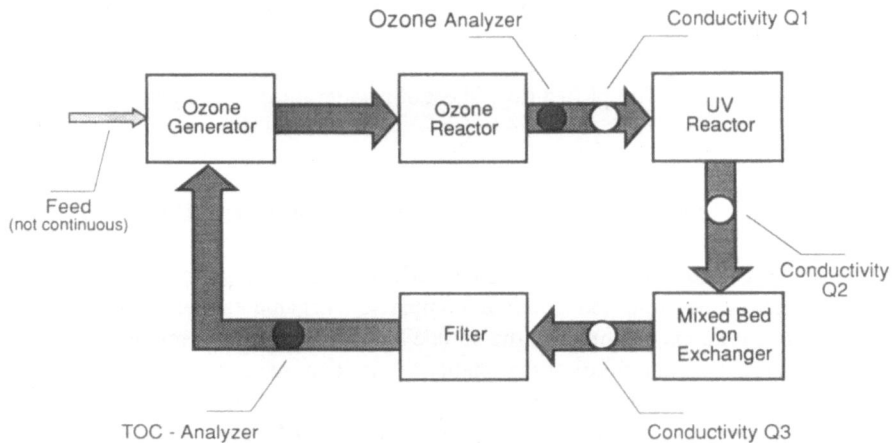

Fig. 2:   Test loop used for laboratory experiments on TOC removal.

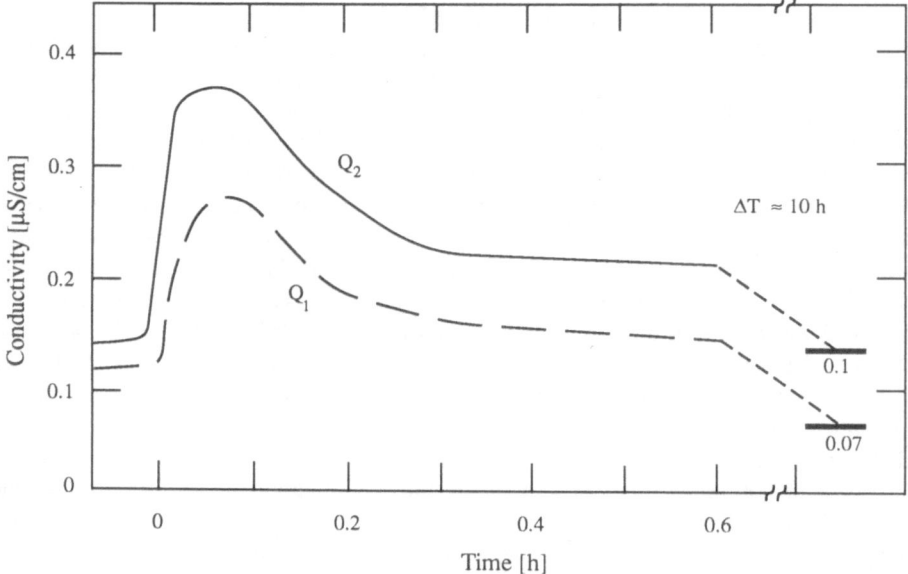

Fig. 3:   Conductivity-transients recorded by the conductivity meters in the laboratory test loop after switching on O_3 and UV. Water as received from DI water supply of the BBC Research Center.

Isopropanol was chosen as the test molecule for the experimental studies. It was used in previous studies[12] on UV/ozone oxidation, because:

1) it is a frequently used solvent in semiconductor processing and hence often found in water recycling from semiconductor rinsing stations;

2) isopropanol does not absorb UV light of 254 nm wavelength and therefore does not react photochemically;

3) the direct reaction of ozone with this molecule is slow enough to be negligible (the half life time of ozone in the presence of isopropanol is 5 h)[10];

4) being undissociated, isopropanol is not absorbed by mixed bed ion-exchanger resins.

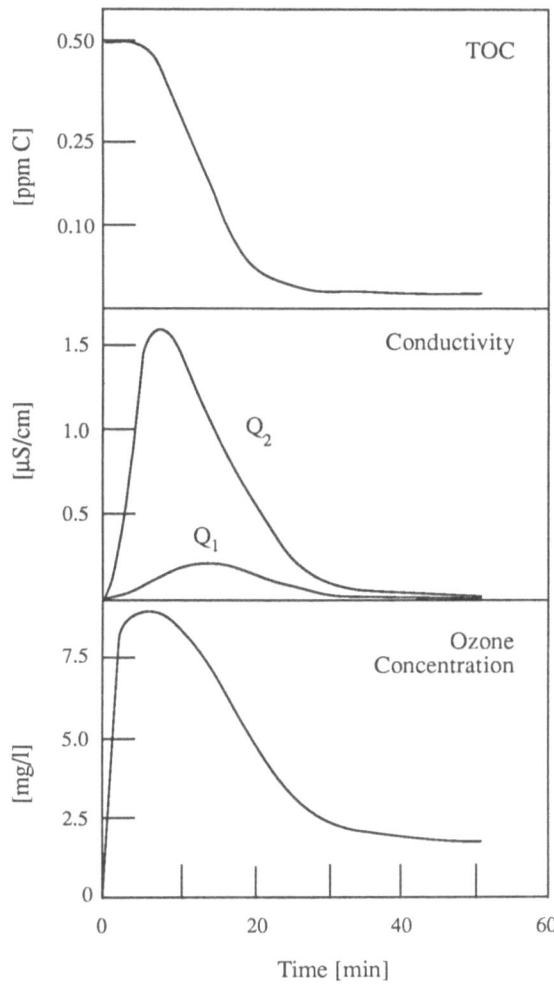

Fig. 4: Transients (conductivity, TOC, ozone concentration) recorded in the laboratory test loop after addition of 500 ppb of TOC in the form of 2–propanol.

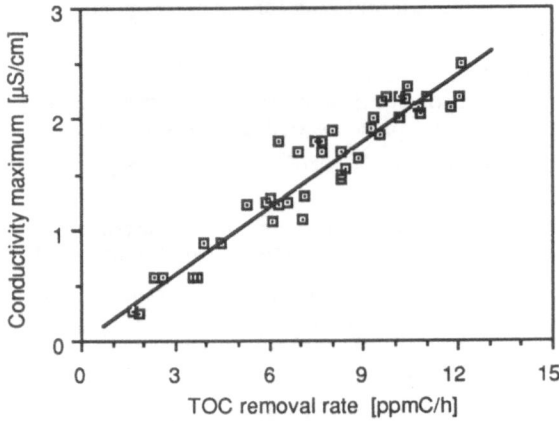

Fig. 5:   Conductivity maximum $Q_{2,max}$ vs. TOC removal rate. Each data point
            corresponds to a 2-propanol injection experiment as shown in Fig. 4.

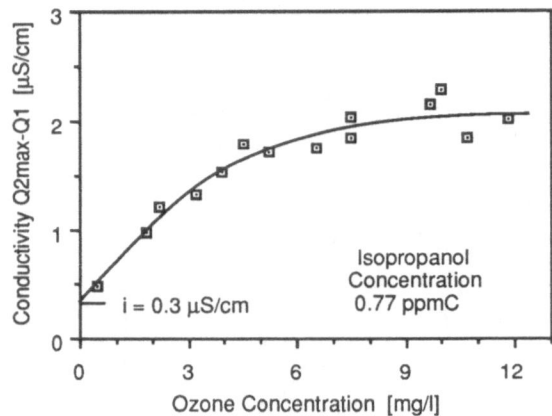

Fig. 6:   Influence of ozone concentration on TOC removal rate (measured as the
            conductivity maximum Q_2 max - Q_1).

4.1. UV/Ozone Process

Fig. 3 shows the conductivity transients measured at the location 1 and 2 of the loop, filled
with demineralized water from the central ultrapure water distribution of the laboratory. At
$t = 0$, ozone dosage and UV lights were switched on, causing a sharp increase in the
conductivities at the outlets of the respective reactors. The transients indicate the presence of
organic contaminations, which are oxidized directly or promoted photochemically. The
system eventually reached a steady state where the water contained virtually zero TOC levels.
At this point, the ozone dosage and UV radiation were stopped and aliquots of isopropanol were
added to the water to give a uniform concentration of isopropanol in the order of 250 to
1000 ppb. Fig. 4 shows the transients recorded with "artificial" TOC (i.e. isopropanol), with

ozone dosage and UV again being started at t = 0. Fig. 4 also shows the recordings of the TOC and ozone analyzers. The conductivity at position 3 was not plotted, but was found to be unaffected by the processes, i.e. the mixed bed was found to be an effective scavenger for the resulting reaction products. While the conductivity after the UV reactor showed a steep rise, the conductivity rise in the ozone reactor was somewhat delayed. The ozone dosage was kept constant (constant output at the Membrel cell). This means that the chemical ozone decomposition rate increased in the course of the experiment. The question is still open, however, whether this increase is due to the accumulation of chain promoters (e.g. hydrogen peroxide) or the destruction of chain inhibitors (such as isopropanol itself) by the process.

The peak value of the overall conductivity increase correlates with the TOC removal rate, which can be evaluated from the slope of the TOC transients. A calibration curve, obtained from a series of experiments, is given in Fig. 5. The plot of conductivity maximum vs. the TOC removal rate approximates a straight line. This means that, in a technical application, simple conductivity measurements could be used to monitor the level of TOC in the water which is being treated.

In order to determine the ozone doses which are required for TOC removal, the influence of the concentrations of ozone and isopropanol was investigated. The other parameters, like flux (i.e. recirculation rate), UV intensity and reactor volumes (i.e. residence time) were kept constant. The influence of ozone concentration at a constant initial concentration of TOC (1 ppm) is shown in Fig. 6. The resulting curve shows that the rate is roughly proportional to the ozone input into the UV reactor up to ~ 5 mg/l of ozone. Higher concentrations do not lead to a significant further increase in the reaction rate, rather the curve seems to reach an asymptotic value. The value of ~ 2 μS cm^{-1} is, of course, specific for the set of parameters that were kept constant. The extrapolation to zero ozone concentration gives the purely photochemical contribution to the oxidation process (in the presence of O_2 saturation).

The rate of TOC oxidation is also a function of TOC level in water. Fig. 7 shows a plot of the conductivity maxima observed as a function of isopropanol concentration. The ozone concentration was kept within 8 - 13 mg/l (flat part of the curve in Fig. 6).

These data enable us to evaluate the efficiency of the UV/ozone process for oxidative TOC

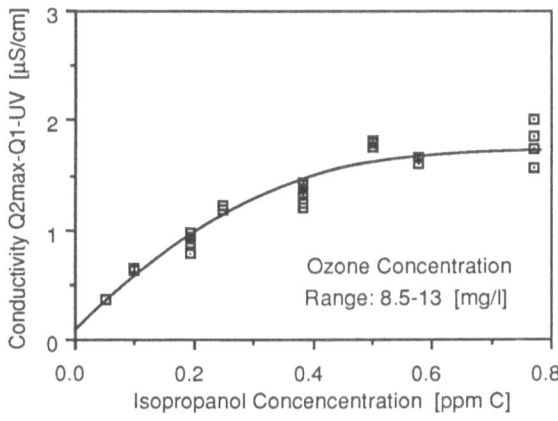

Fig. 7:   Influence of 2-propanol concentration on TOC removal rate (measured as the conductivity maximum Q_2 max - Q_1).

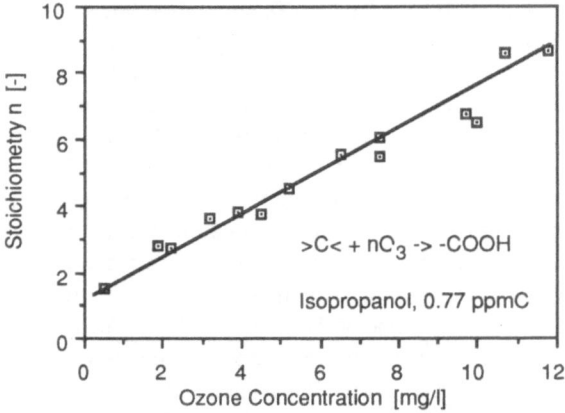

Fig. 8:   Stoichiometry of TOC removal with 2-propanol as test molecule. The
            coefficient n indicates the number of O_3 moles required to remove one
            mole of organic carbon.

removal in ultrapure water and to estimate the technical feasibility of the process. Fig. 8 shows
that the number n of moles of ozone needed for the elimination of one mole of organic carbon
increases with the ozone concentration employed. The process needs to be optimized with
respect to efficiency as well as reaction rate (i.e. residence time). From the curve in Fig. 6
(dependence of reaction rate on ozone concentration) we would expect the optimum ozone
concentration to be in the order of 5 mg/l for 1 ppm of TOC. This gives an oxidation
stoichiometry of n ~ 4 in an overall reaction scheme:

$$C + nO_3 \rightarrow -COOH .$$

It is obvious that n will assume different values for other molecules or "unknown TOC".
Assuming a range of 2   n   10, we can estimate the ozone demand for TOC elimination in an
ultrapure water loop. For the elimination of an initial TOC of 200 ppb, the ozone demand would
be in the order of 2 to 7g/m^3. Higher concentrations of TOC (1 - 10 ppm) are found in water
reclamation systems, i.e. in the water which is recycled from semiconductor rinsing stations.
This water typically contains low molecular weight TOC impurities like isopropanol. For such
applications, ozone doses of up to 100 gm^{-3} would be necessary.

4.2. Hydrogen peroxide/ozone process

According to the literature[13,14,15], the photolysis of ozone in water yields hydrogen peroxide
as the first intermediate, which then acts as a chain promoter for the chemical decomposition
of ozone and the formation of OH radicals. If the photolysis of ozone exclusively leads to H_2O_2,
it should be possible to achieve the same oxidation power by dosing hydrogen peroxide instead
of using UV irradiation. In order to test this assumption, experiments were run with a constant
H_2O_2 and O_3 feed to the experimental loop Fig. 2. The rate of isopropanol elimination was
measured via the conductivity meters for a series of experiments with H_2O_2 dosage as a
parameter.

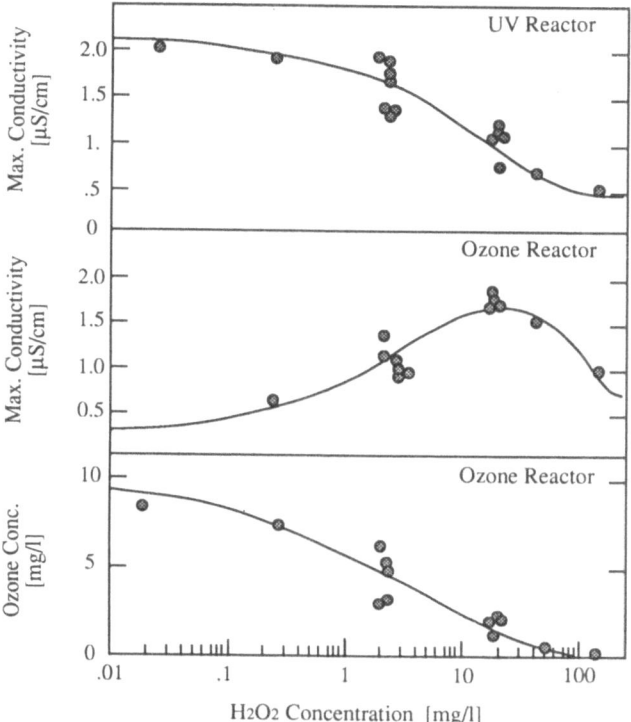

Fig. 9: Ozone/hydrogen peroxide process

Top: TOC removal rate in UV reactor (measured as the conductivity maximum Q_2 max - Q_1).

Middle: TOC removal rate in chemical reactor (measured as the conductivity maximum Q_1).

Bottom: Ozone concentration at the outlet of the chemical reactor. All as a function of H_2O_2 concentration in the feed. The amounts of 2–propanol additions and the ozone dose were kept constant.

The conductivity changes (peak values at the reactor outlet, corrected by the value at the reactor inlet) together with the ozone concentration at the outlet of the chemical reactor are plotted as a function of H_2O_2 concentration in Fig. 6. The ozone dose (i.e. the production rate of the Membrel cell) was kept constant at a value in the order of 15 mg/l.

The oxidation reaction in the H_2O_2/O_3 reactor proceeds with the fastest rate (maximum of curve b) if comparable doses of H_2O_2 and ozone are used[14]. The maximum of curve b is comparable with the maximum rate observed in the photochemical reactor at vanishing H_2O_2 dosage. The results indicate that the assumption of H_2O_2 as an intermediate in photochemical ozone decomposition is probably valid[15].

Although the TOC removal rates obtainable by H_2O_2/ozone are comparable to those observed under the same conditions by UV/ozone, peroxide dosage does not seem to be a viable process in ultrapure water mainly for two reasons:

a) Introducing another chemical into the system with the inherent danger of ultrapure water contamination. This is particularly the case with H_2O_2 solutions which are known to contain stabilizers.

b) Massive overdoses would be necessary to guarantee complete removal of ozone[15].

One point which remains to be discussed is the influence of hydrogen peroxide, whether added intentionally or by photochemical conversion from ozone to the ultrapure water system. Studies by Martinola[16] have shown that common mixed-bed ion-exchangers are capable of removing hydrogen peroxide. The anion exchanger resins catalytically decompose H_2O_2. Their performance is, however, not stable and breakthrough through the mixed-bed polisher of H_2O_2 cannot be excluded after prolonged operation of the polisher with H_2O_2 containing water. The use of noble metal catalysts, supported on ion-exchange resin beads, will have to be tested for trace H_2O_2 elimination.

5. Conclusions and Outlook

Membrel ozone generators have found widespread applications in the disinfection of ultrapure water systems. The doses necessary to maintain low bacterial counts in ultrapure water are very low (0.02 gm⁻³). The combination of UV with ozone dosage allows efficient oxidation of trace organic water contaminants to carboxylic acids which can be removed by the ion-exchange polisher.

The UV/ozone process needs ozone dosages in the order of 1 to 100 g/m³, depending on the TOC content of the water. The ozone generating capacity for this process has to be several orders of magnitude higher than for ultrapure water disinfection.

References

1   Dhawan, G.K. and Martinola, F., TOC and Low Pressure Reverse Osmosis for Final Water Treatment, WSIA 12th Annual Conf., Orlando, Fla. 13. - 18.5. 1984, Proc. Vol. 1.

2   McPherson, H., Ultrapure Water, Jan/Feb (1986), 20.

3   Setz, W., Pharm. Ind. **47** (1985) 3.

4   Vogel, L. and Klein, H.P., Brown Boveri Review **8** (1986) 451.

5   Stucki, S., Theis, J., Kötz, R., Devantay, H. and Christen, H.J., J. Electrochem. Soc. **132** (1985) 367.

6   Baumann, H. and Stucki, S., Swiss Chem **8**, 10a (1986) 31.

7   Stucki, S., Baumann, H., Christen, H.J. and Kötz, R., J. Applied Electrochem. **17** (1987) 773.

8   Poirier, S.J. and Kantor, K.J., TOC reduction. Ultrapure Water . July/Aug. (1987) 40-45

9   Kosaka, K., Yokoyama, F., Koike, K. and.Urai, N., Proc. Ultrapurewater Journal's 1st High Purity Water Conf., April 1987, Philadelphia, p. 143.

10 Hoigné, J. and Bader, H., Water Res. **17** (1983) 173.

11 Baumann, H. and Stucki, S., Proc. 8th Ozone World Congress, Sept. 1987, Zürich. International Ozone Association (IOA). (Unionsverlag, Zürich) 1987, Vol. 2, pp. K12-25.

12 Barker, R. and Taylor, F.M., Proc. Int. Conf. on the Role of Ozone in Water and Wastewater Treatment, London, Nov. 1985, pp. 106-116.

13 Taube, H., Trans. Faraday Soc. **53** (1957) 656.

14 Glaze, W.H., Kay, J.-W. and Aieta, M., Proc. 2nd Int. Conf. on "The Role of Ozone in Water and Wastewater Treatment", April 1987, p. 233.

15 Hoigné, J. and Bader, H., Proc. 8th Ozone World Congress, Sept. 1987, Zürich. , Internationaö Ozone Association (IOA). (Unionsverlag, Zürich) 1987, Vol. 2, pp. K83-97.

16 Martinola, F., VGB Kraftwerkstechnik **58**, 6 (1978) 1.

Discussion

Chairman: A. Miquel
Brown Boveri, Baden, Switzerland

P. Francis   •   Where, in your recirculating loop (Fig. 2), would you consider the point of use to be in order to preserve the minimum amount of TOC? You have got two problems: one is not to have ozone in the water and the other is to make sure that there is no TOC leached from your mixed bed.

S. Stucki   •   Our simulated point of use was the TOC analyzer. At this point we have no conductivity - 18 megohms, we reach near zero levels of TOC and we certainly have no ozone because it is all destroyed by the UV reactor.

C. von Sonntag   •   Do you have at this point any hydrogen peroxide left (i.e. at the TOC analyzer point)?

S. Stucki   •   If so, its concentration is below what we can measure at present. We can only estimate an upper limit from our measurements with H_2O_2 additions. The estimated limit is ≤ 0.1 ppm.

W. McRae   •   I just wanted to ask when you added hydrogen peroxide did you also use UV?

S. Stucki   •   Yes, just to make sure that ozone does not get into the mixed bed.

K. Wieck-Hansen   •   All your trials had been made with isopropanol as far as I could see. I would like to know if you have had any experience with very complex molecules. I am thinking of humic acid or the like.

S. Stucki   •   We have concentrated on isopropanol. Some experiments have been done with acetone which exhibits exactly the same removal rates, for obvious reasons as Prof. Hoigné would say. The only thing I can refer you to is the initial transient we measure with ultrapure water taken from the central ultrapure water supply of the laboratory. We know there is some residual TOC in that water, but we don't know what it is. Since the oxidation reaction of ozone proceeds, in our case via hydroxyl radicals, one would not expect the reaction to be very specific for different molecules. We are currently investigating other molecules like halomethanes, but I can't really comment on these results yet.

J. Hoigné   •   I think isopropanol is a fair test molecule because other products are also first degraded to such types of molecule. A comment on the different lifetimes of the ozone in the chemical reactor: they are rather typical. It is known that traces of propanol stabilize aqueous solutions of ozone. In the absence of these traces of isopropanol, ozone decays via a non-inhibited, radical type chain reaction. Since in your system there is no inhibitor left, your curve really corresponds to what is expected from reaction kinetics.

S. Stucki   •   Thank you for that comment. We do think that there is inhibition by the trace TOC itself. If the pure water is really pure, the ozone decay kinetics seems to become more and more difficult to control. That means that all the data in the literature on ozone stability in pure water is questionable with respect to the presence or absence of trace impurities.

Treatment Techniques for Waste Water from Chemical Industries

E. Plattner and Ch. Comninellis
EPFL, Lausanne, Switzerland

1. Introduction

Besides inorganic materials (acids, bases and salts), industrial waste water also contains organic substances which have to be "mineralized", i.e. oxidized to CO_2, H_2O, N_2 (or NO_3^-), Cl^- and SO_4^{--}, before the water can be discharged. The main methods for analyzing organic material in waste water are :

TOC	:	total organic carbon	[mg C/l]
COD	:	chemical oxygen demand	[mg O/l]
BOD_5	:	biological oxygen demand in five days	[mg O/l].

The first two methods determine the total amount of organics. Depending on the composition, the relationship COD/TOC varies between 0.67 (oxalic acid) and 5.33 (methane). Typical values for aromatic compounds lie between 2.7 and 3.3. The ratio BOD_5/COD indicates the fraction of organic materials which are biologically degradable. The difference (COD-BOD_5) is said to be refractory.

Today, the tendency is for legislation to prefer to limit the daily discharge of pollutants rather than setting restrictions on permissible concentrations in waste water.

The different treatment techniques can be divided into two groups:

* biological and/or chemical oxidation techniques which really "clean" the water;
* separation and concentration techniques allowing the pollutants to pass from one phase to another. Recovery and recycling are, however, possible.

2. Biological and Chemical Processes

These processes can be divided according to the oxidation temperature (Table 1). Up to approximately 300 °C (under pressure) the reactions take place in the liquid phase.

Above this temperature, oxidation reactions proceed at high speed with complete elimination of TOC. Salt separation is possible. The energy consumption per m³ of waste water increases with the temperature (see Fig. 1).

Table 1: Oxidation processes and reaction temperature.

Reaction Temperature	Oxidant	Process	TOC Reduction	Comment
20 - 40 °C	air/O_2	biology	partial	cheap
20 - 90 °C	Cl_2/ClO_2			
"	H_2O_2	chemistry	partial	
"	ozone		to	expensive
"	electric current	electrochemistry	total	
200-300 °C	air/O_2	wet oxidation	total	100-200 bar
400-550 °C	air/O_2	supercritical	total	> 220 bar
550-620 °C	air	calcination	total	
750-850 °C	"	salt melt	total	
> 1000 °C	"	combustion	total	

Table 2: Influence of Substituents on the Oxidizability of Benzene Derivatives[3].

Substituents			Chlorine Consumption[*] mole Cl_2 /mole substrate
o-	m-	p-	
-OH			9.8
-NH$_2$			8.3
-NO$_2$			0.1
-COOH			0.3
-NO$_2$	-NH$_2$		8.5
-NO$_2$	-COOH		0.1
-COOH		-Cl	0.1

*) 15 h at room temperature, pH 7 and 20 mol Cl_2 per mole of substrate. Initial substrate concentration: 10^{-4} mole/l.

2.1. Low Temperature Processes

Biological purification is the most common technique and is the final step in water treatment, but it does not meet, on its own, the set requirements for chemical waste water. This means that more powerful and expensive techniques have to be used before biological treatment. Aims for development are:

• improvement of the space/time yield through higher temperatures, better mass transfer (use of oxygen-enriched air or pressure), immobilized biomass, etc.;

• improvement of TOC elimination by employing product-specific bacteria, etc.;

• anaerobic biology (methane-recovery).

Oxidation with chemicals or electricity at temperatures up to 100 °C is expensive. It is employed in specific cases where other techniques fail (for example, in very diluted, refractory waste water).

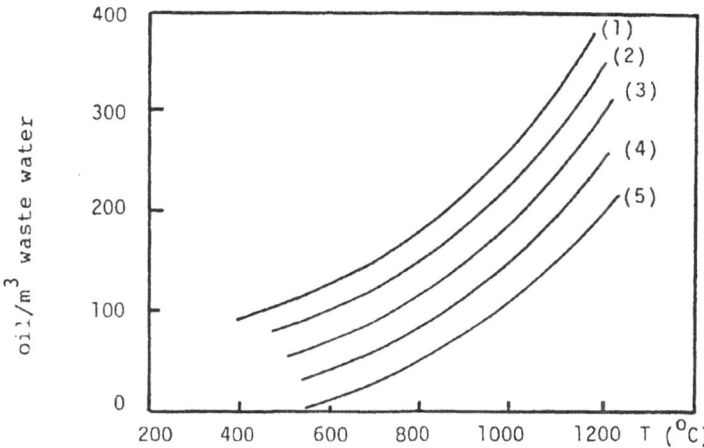

Fig. 1;   Approximate energy consumption for various oxidation temperatures and heating values of the waste water.

(1)	0	(GJ/m^3)	(2)	1.2	(GJ/m^3)	(3)	2.4 (GJ/m^3)
(4)	3.6	(GJ/m^3)	(5)	5.0	(GJ/m^3)		

Chlorine, sodium hypochlorite and chlorine dioxide are usually used as oxidants. Chlorine dioxide hydrolyses to chloric and chlorous acids.

$$2\ ClO_2 + H_2O \quad \rightarrow \quad HClO_3 + HClO_2 .$$

Only the latter acts as oxidant ($E_o = 1.64$ V) since chlorate ions are stable in water. Chlorine dioxide is not commonly used in waste water treatment.

Chlorine or sodium hypochlorite (Javelle-water) act best in slightly acid waste water ($E_o = 1.49$ V) where they appear as undissociated hypochloric acids. In more acidic conditions, chlorine is evolved and in basic solutions hypochlorite ions are less powerful oxidants ($E_o = 0.90$ V).

Hypochloric acid acts as an oxidizing and chlorinating agent for organic substances :

Reaction 1

This property is a disadvantage in waste water treatment since chlorinated derivatives are generally toxic and difficult to decompose. For example, phenol is chlorinated several times before oxidation into chloromaleic and trichloroacetic acid occurs [1,2].

The effect of substituents on the oxidizability of the benzene derivatives is shown in Table 2.

Activating substituents (-OH, -NH$_2$, -NR$_1$R$_2$) promote the reaction considerably, even in the presence of otherwise stabilizing or inhibiting substituents (-NO$_2$, -SO$_3$H, -COOH, -Cl,...). The same effect has been observed with the use of other oxidants as well as with electrochemical oxidation[3].

Hydrogen peroxide shows a higher oxidation potential (in slightly acidic environments) than chlorous acid (E$_o$ = 1.77 V). Oxidation takes place in the presence of iron as a catalyst and often results in oxalic acid as an end product. The reaction products are generally non-toxic and biologically degradable. The ecological advantages of hydrogen peroxide in comparison with chlorine have to be set against the higher costs involved. The treatment technique is very simple and can be carried out in a conventional apparatus (stirred reactor). The iron catalyst is precipitated out at the end of the reaction and remains in the tank for the next batch.

Ozone is the strongest oxidant in the group (E$_o$ = 2.07 V) and attacks practically all organic products. Phenol oxidation has been investigated by several authors[3].

Reaction 2

The main end product of oxidation is oxalic acid. It is also to be found analytically after oxidation of real waste water (with nitrated and sulphonated benzene derivatives) and is reflected by the relationship COD/TOC < 1. The specific ozone consumption (gO$_3$/gCOD) increases sharply during the oxidation (Fig. 2) which means decreased oxidation efficiency.

Ozone is produced in air or oxygen, but can only react in dissolved form, which involves a mass-transfer from the gas to the liquid phase . Relatively large extraction columns as well as an off-gas treatment are necessary. For plant design, optimization of the following parameters is required: pH, partial pressure of ozone and specific contact areas. The costs of ozonation (investment and operating costs) are high. The process is only economically applicable with very diluted, refractory waste water.

The electrochemical anode potential in aqueous media is limited by the formation of oxygen (E$_o$ = 0.815 V at pH 7). Using electrodes with high oxygen overvoltage, this potential can be raised to 1.4 V (Pt) or 1.6 V (PbO$_2$), i.e. to an oxidation potential between that of chlorine and hydrogen peroxide.

The anode has to be stable in different waste waters and still remain catalytically active for oxidation. These conditions are not easy to fulfill and this process is not yet available on an industrial scale.

Electrochemical oxidation, like ozonation, is considerably faster in alkaline solutions.

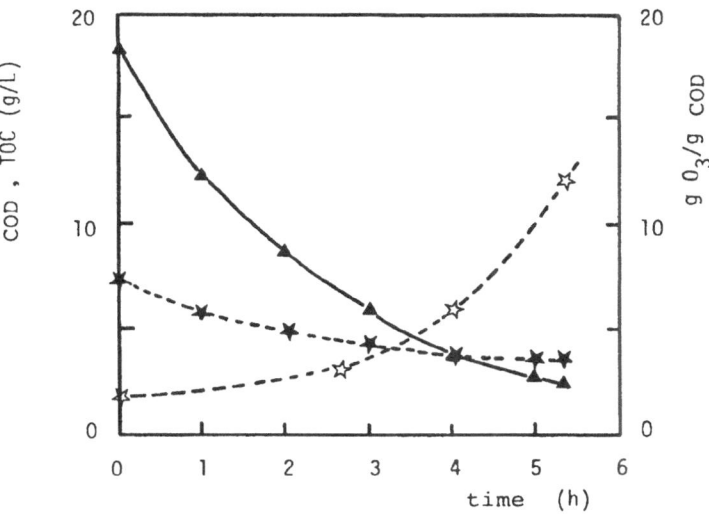

Fig. 2:   Waste water treatment using ozone; COD, TOC and ozone consumption of
waste water; 60 *l*/h of oxygen with 45 mg/*l* ozone, temperature 28 °C;
pH = 13, constant.
▲ COD
★ TOC
☆ specific ozone consumption (g O₃/g COD)

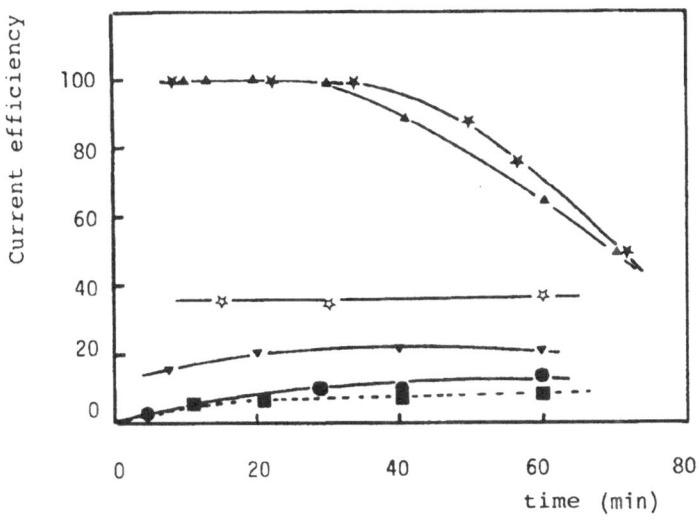

Fig. 3:   Current efficiency for the electrochemical oxidation of various derivatives
of benzene:
★ p-ATS   ▼ 4-nitrotoluene-2-sulfonic acid
▲ aniline   ● benzene sulfonic acid
☆ phenol   ■ benzoic acid

Further oxidation steps of aromatic derivatives leads to the formation of maleic acid (and not oxalic acid as in ozonation), which means less oxidant consumption.

Substituents also have considerable influence on oxidizability whereby the presence of a single activating group is sufficient to produce a high current efficiency (Fig.3). This explains a further advantage of electrochemical treatment, which can be used both for reduction and oxidation (conversion of the stabilizing -NO_2 into an activating -NH_2 group). Current or overall efficiency in alkaline media depends on the nature of the substances and their degree of elimination. The loss in current efficiency corresponds to oxygen formation. After 120 min. of electrolysis, 70% of initial COD and 48% of TOC are eliminated and 90% of p-ATS appear as maleic acid (Fig.4). The current efficiency of oxidation calculated from the composition of the solution is about 58%. This accords well with oxygen formation measured during the experiment (Fig. 5).

With electricity consumption of approximately 6 kAh per kg COD and 3.5 V cell voltage, 91.4 kA or 320 kW are employed for the treatment of 10 kg/h p-ATS in waste water.

2.2. Comparison of the Chemical and Electrochemical Techniques at Low Temperature

The price of the oxidation equivalent varies. Oxygen in air is the cheapest, followed by chlorine, then electricity, hydrogen peroxide and finally ozone (Table 3). Oxidation with oxygen at low temperature is only possible biologically. Chlorine often forms non-decomposable and toxic chlorinated derivatives. This restricts its application.

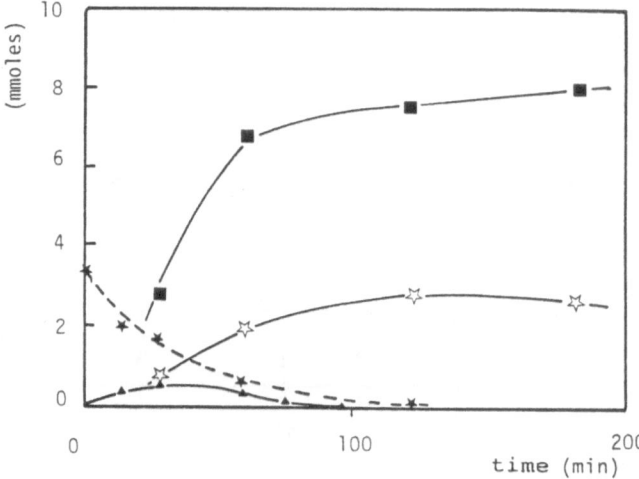

Fig. 4:   Intermediates in the electrochemical oxidation of 4-aminotoluene-2-sulfonic-acid (p-ATS).
Electrolyte: 15% Na_2SO_4, 0.4% NaOH, $2.3.10^{-2}$ M p-ATS
Current density: 60 mA/cm², temperature 70 °C, Pt anode, volume 150 ml
★   p-ATS   ☆   substituted maleic acid
▲   substituted quinone   ■ CO_2

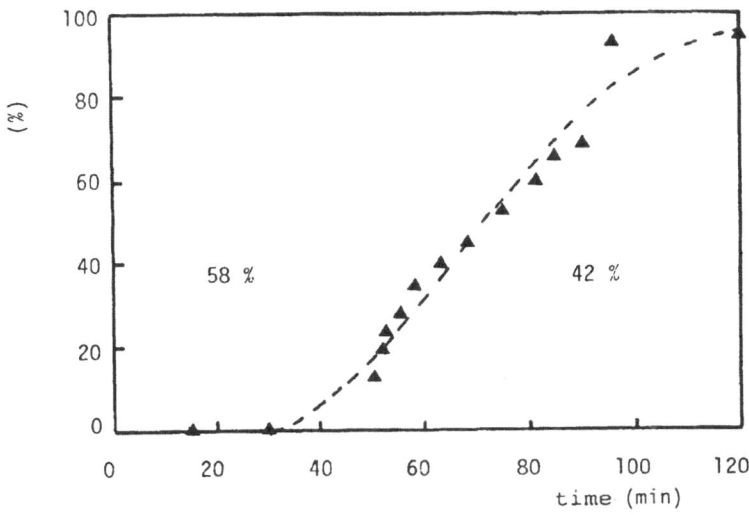

Fig. 5:   Oxygen formation (% of the total current) in the oxidation of p-ATS.

Table 3:   Comparison of the various (electro-) chemical processes.

Oxidant	SFr./kg equiv.	Average Yield	Invest. Cost for 100 kg COD/h (M SFr.)	Treatment Cost SFr./kg COD
O_2*	< 1.-		2 – 6.0	1 – 3
Cl	18.-	50%	3.0	6.50
H_2O_2	45.-	55%	3.0	12.-
O_3**	60.-	100%	24.0	14.-
Kwh	10.-	56%	10.0	6.10

*)   Biological treatment
**)   Only two equivalents / mole

Electrochemical oxidation is clean, and, in the presence of activating substituents, efficient. Its use is still limited by lack of stability of the electrodes. Hydrogen peroxide can be used as an ad hoc but expensive oxidant where very small quantities of waste water have to be detoxified. Ozone is the most expensive, but also the most powerful, oxidant which can be used in all situations. The investment costs for production and treatment plants are highest.

2.3. Processes at Elevated Temperatures

Wet oxidation, or combustion in the liquid phase with molecular oxygen at 200 - 300 °C and 50 - 200 bars, is a powerful and practically quantitative process. Within 1/2 to 2 hours in the presence of a Cu-catalyst, almost total elimination of the TOC is possible, even with refractory products such as dioxin, diphenyl, machine oil, pyridin, etc.[5].

Organically bound C and H are oxidized to CO_2 and H_2O (with traces of CO). -NO_2 is converted to N_2, -NH_2 to NH_3 and organic Cl, S, P to HCl, H_2SO_4 and H_3PO_4, respectively.

A full scale plant (Fig. 6) consists essentially of: reactor (bubble columns up to 30 m³ in volume), heat exchanger, high pressure pump and compressor. Waste water is conditioned in the collecting tank, then mixed with the catalyst and pumped into the reactor via the heat exchanger (10 -50 m³/h). Air (10 % excess) is compressed to 50 - 200 bars and fed into the waste water. After cooling, the oxidized mixture is separated into gas and liquid phases. The catalyst is precipitated, filtered out and recycled. Ammonia is reclaimed and carbon monoxide is burnt. Numerous technical problems had to be solved (e.g. corrosion and scaling of heat exchanger and reactor) before the first large plant could be put into operation. Investment is high but, because the process is specially suited to handling heavily contaminated waste water (approximately 100 g COD/*l*), both investment and treatment costs (4.0 mio SFr. per 100 kg COD/h and approximately 3.0 SFr. per kg COD) compete with other processes.

The *Modar-process* (or combustion in super-critical steam) has not, to our knowledge, surmounted scaling-up hurdles. Oxidation takes place in a homogeneous phase at higher temperatures than wet oxidation (approx. 550 °C) and is completed within minutes. In

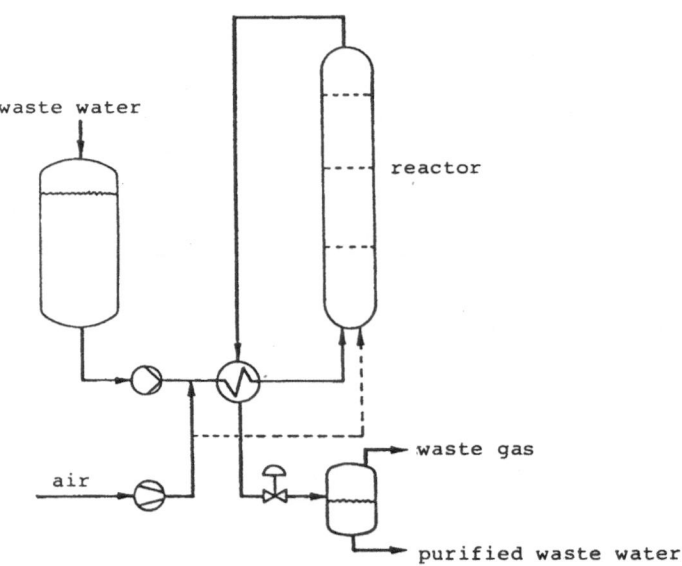

reaction conditions: T = 200 - 300 °C
                      P = 50 - 200 bar

Fig. 6:   Wet oxidation: plant flow chart.

principle, this process allows the separation of inorganic salts, either in solid form (sulphate or phosphate) or as highly concentrated solutions (chlorides). In order to avoid scaling, the input water is directly heated with part of the super-critical steam escaping from the salt separator (Fig 7)[4].

With over 5% organic products in the waste water, the system is said to be energy self-sufficient. The water so produced is clean, and this process has therefore even been proposed as an alternative to desalination of seawater through direct addition of fuel.

Oxidation in molten salts by blowing through air is only possible with the solid residue of waste water evaporation. The organics must not be volatile. Waste water, for example from letter acid production (naphtaline sulphonic acid derivatives), would be suitable for such a treatment. In contrast with pure combustion or pyrolysis at temperatures of 1000 - 1200 °C, oxidation in molten salts at 750 - 850 °C has the great advantage of producing smaller amounts of practically salt-free flue gases.

Waste water combustion has often been used for refractory and poisonous waste water even though investment costs (3 - 5 mio. SFr./m^{3}h), energy consumption ($\sim$ 0.2 tons heavy oil/m^3), and operating costs (200 - 300 SFr./m^3) are high.

Pre-treated waste water is pyrolized in a furnace at about 1000 °C with excess air. The hot waste gases (approx. 11 tons per m^3 waste water) are first cooled in a boiler, then in a tower by quenching with water and are freed from accompaning dust by electrofilters.

The relatively complicated plant (Fig. 8) requires intensive maintenance. Its availability is seldom more than 70% (due to corrosion on the furnace walls and metal parts, plugging by salt deposits, etc.). A particular disadvantage is the large volume of flue gas produced which contains, besides water and carbon dioxide, some SO$_2$ and NO$_x$.

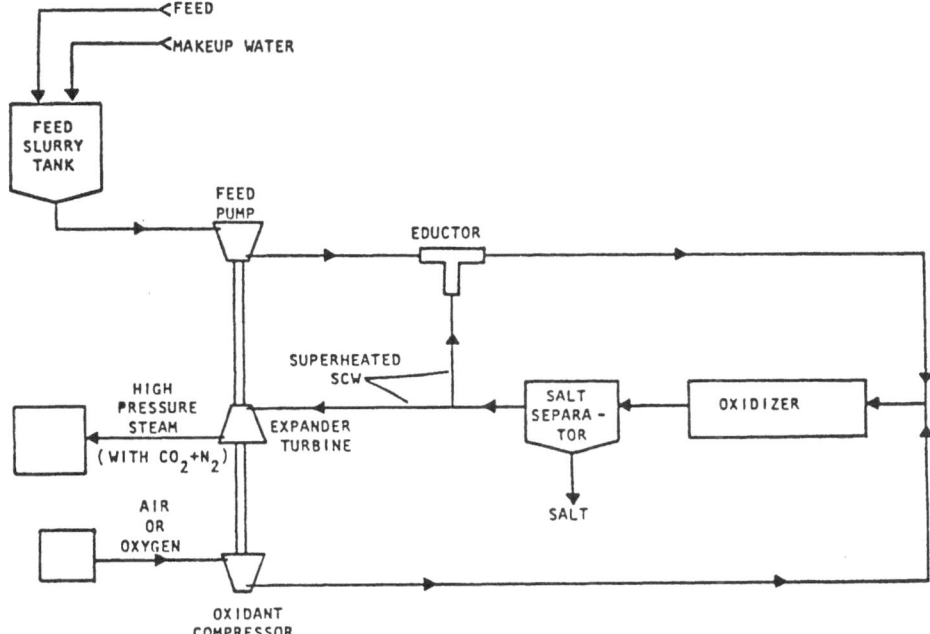

Fig. 7:   Flow chart of the MODAR process for oxidations in super-critical environment.

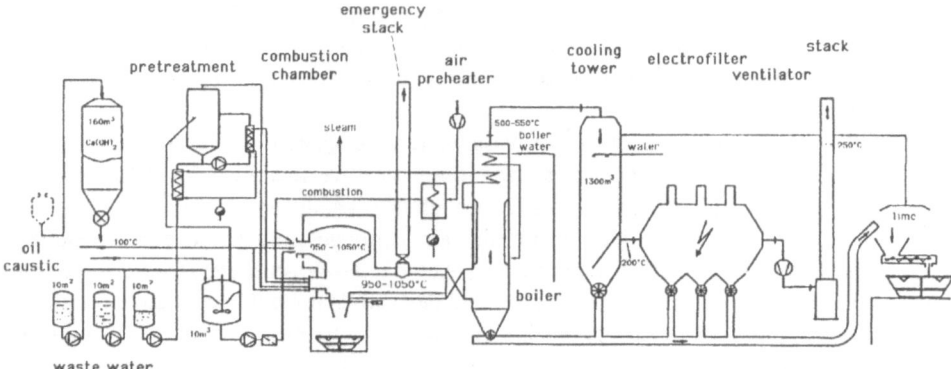

Fig. 8:   Waste water combustion plant of CIBA-GEIGY, Monthey, building 388A.

Babcock has developed a pyrolysis process with post-combustion for processing waste and old rubber which should present an improvement. Waste is pyrolized in the presence of calcium oxide at 600 °C in a rotating furnace. The furnace is heated indirectly by the post combustion of the pyrolysis gases. Inorganic salts and any coke present accumulate as slag.

Investment and operating costs are lower than in direct combustion since the furnace and solid matter are only heated to 600 °C. The quantity of waste gas is 40% smaller. The Babcock–process has yet to be tried out with concentrated waste water.

3. Separation and Concentration Processes

Waste water is often produced in relatively diluted form. In high temperature oxidation processes (> 300°C), it makes sense to concentrate the waste water first, so that the whole process operates more efficiently. Possible separation processes can be divided into two groups according to whether organic products or "clean" water is separated from the waste water.

3.1. Isolation of Organic Components

These processes allow in principle the reclaiming and recycling of the separate organic products.

Adsorption relies on different retention between dissolved organic molecules and the inorganic components of the waste water on the surface of the adsorption agent. Activated carbon, the most common agent, is capable of binding up to a third of its weight in organic material before these substances appear in the drained off water. Used carbon can either be burnt with organic products (small plants), or it can be regenerated by pyrolysis or steam treatment.

Adsorption plants are very effective and unproblematic. Regeneration plants, on the other hand, are complicated and expensive (4 - 6 SFr./kg activated carbon or 5 - 10 SFr./kg COD). The application of this process is restricted to special cases, e.g. very diluted poisonous waste water.

Extraction makes use of the difference in solubility of the products to be separated in water and a water insoluble solvent. The latter flows against the current of waste water in a column. The degree of extraction depends on, among other things, the number of theoretical stages, i.e. the height of the column.

The solvent carrying the separated matter has to be regenerated using distillation or re-extraction. The organic product accumulates in concentrated form and can either be recycled or destroyed through combustion.

Of special interest is the use of liquid ion-exchangers such as trialkylamine (C_8 - C_{10} alkyl substituents) to extract carboxylic and sulphonic acids from the waste water. TOC removal over 80% has been achieved for real waste water. Re-extraction with caustic soda solution results in a concentration factor of over 10. Extraction costs, excluding concentration treatments, are in the order of 50 SFr./m³ or 1 - 3 SFr./kg COD.

Stripping with a gas or preferably with steam (because total condensation is possible) is the favoured method for removing solvents or volatile organic substances from waste water. The solvents, which are normally separated by decanting the condensate, can either be recycled or burnt. Costs are in the order of 10 - 20 SFr./m³ or 0.5 - 2 SFr./kg COD.

3.2. Separation of Water

The following processes concentrate the waste water. The resulting separated water is either recycled or discarded. The concentrate has to be mineralized through oxidation.

Membrane processes have been part of waste water treatment for several years now. Although these processes are, in principle, very suitable (the waste water is separated into permeate and concentrate without phase change), they are limited by such matters as:

- osmotic pressure (reverse osmosis);
- limit of retention (separation salt/organic material);
- stability with respect to extreme solvents, pH, temperature, pressure;
- fouling (reduction of the permeate flux);
- specificity of the process (membrane suitability has to be tested for each kind of waste water).

Progress is being made in tackling these problems through improvements in membrane quality.

Fig. 9 shows a typical separation plant. Waste water is circulated under pressure over membrane modules and the permeate is continually removed. After approx. 20 hours operation, the initial volume is reduced by 1/10 to 1/20, and the permeate efficiency decreases from 25 to 15 l/m^2 h. After each batch, the plant must be cleaned by rinsing with a complexing solution (danger of scaling). Average retention values vary with the molecular weight of the dissolved substances and with the quality of the membrane. Installed membranes cost about 0.75 mio. SFr. per 100 m², and total operating costs are 10 - 25 SFr./m³ or 1 SFr./kg COD. Average separation yield values for this type of plant vary with the molecular weight of the dissolved substances and with the quality of the membrane:

Intermediate products	(120–200)	:	60 - 70%
Dyes	(> 300)	:	99%
Inorganic Salts	(Cl^-)	:	~ 5%
	(SO_4^{--})	:	~ 35%

Distillation. The application of the well known "process of evaporation" of waste water can be hampered by contamination of the distillate (with volatile products other than water) or salt separation.

Multistage or thermo-compression plants, possibly in combination with salt separation, are used in preference to simple evaporation. With concentration by a factor of 10, the operating costs amount to 40 - 50 SFr./m³ in a 2-stage plant or ≈ 0.5 - 1 SFr./kg COD. The concentrate still has to be disposed of.

Drying is avoided wherever possible because it involves handling of solid material. If necessary, standard procedures and equipment for drying are used with corresponding costs (0.5 - 2 SFr./kg evaporated water).

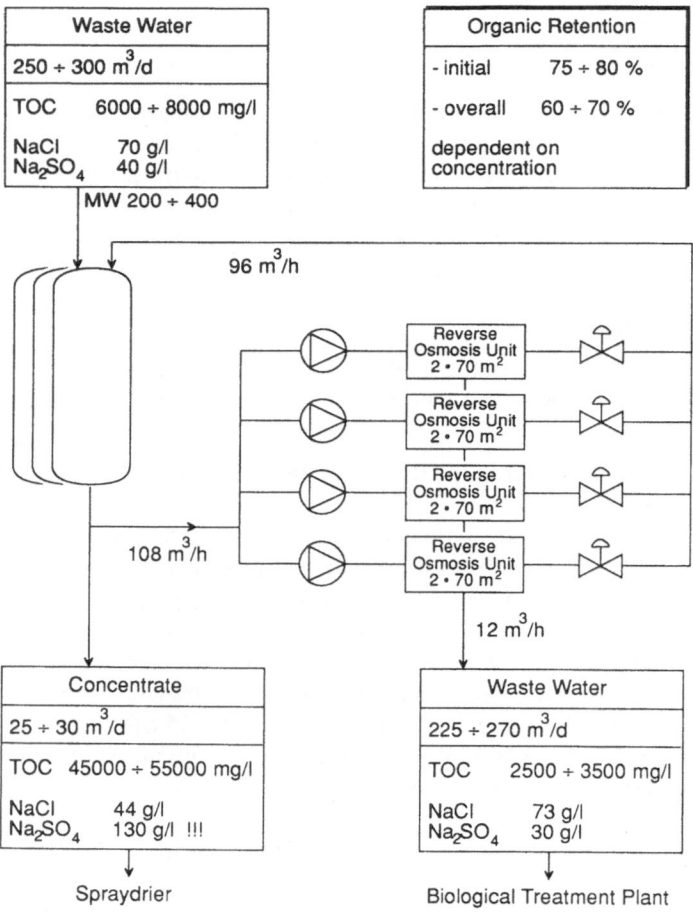

Fig. 9:   Flow chart of waste water concentration by membrane process.

4. Conclusions and Trends in the Purification of Chemical Waste Water

Biological purification, wet oxidation and waste water combustion are three important detoxification processes for chemical waste water. The first is considered a cheap process, the other two expensive. The comparison is valid if the processes are assessed on the basis of their hydraulic load (3 - 10 SFr./m^3 for biological purification, 150 - 200 SFr./m^3 for wet oxidation and 200 - 300 SFr./m^3 for combustion). A comparison of the effective elimination, calculated as kg TOC or kg COD, gives roughly the same costs, 3 SFr./kg COD. This is easily explained when one takes into account the large differences in TOC or COD-content of the waste water to be treated. Biological purification is not suitable for waste water with high or refractory organic load as this can be burnt efficiently. On the other hand, combustion cannot be used for dilute waste water treatment.

If combustion is to be applied, the plant waste water must first be separated into biologically suitable and unsuitable water. Concentration processes, especially those using membranes, can be useful pre-treatments for the second kind of process.

Chemical and electrochemical processes, which are not yet widely employed, are considerably more expensive per kg of decomposed TOC or COD. The combination of ozone-oxidation or better, electrochemical treatment, with biological purification looks promising. Both these more expensive processes can be seen as pre-treatments and the degree of oxidation, i.e. the ozone or electricity consumption, is chosen so as to be just large enough to crack the refractory products into biodegradable smaller molecules (e.g. maleic acid derivatives). This means that, in some circumstances, two separated waste water disposal systems can be avoided if the pre-treatment plant is directly linked to the production line.

Biological purification has not yet been completely optimized. An increase in space/time yield would make this process more economic. An improvement in mass transfer (by better gas dispersion and the application of oxygen-enriched air) has not only increased the through-put, but more importantly, the elimination of TOC (values as high as 85% of TOC have been achieved). The reason for this is possibly that, in this process, the bacteria become more quickly and better adapted to chemicals. If such results can be generalized, then biological methods will have a better future than chemical ones.

References

1 Smith, J.G., Siow-Fong, L. and Netzer, A., Model studies in aqueous chlorination: the chlorination of phenols in dilute aqueous solution. Wat. Res. **10** (1976) 985-990.

2 Smith, J.G. and Siow-Fong, L., Model studies in aqueous chlorination: the chlorination of phenols in dilute aqueous solution. J. Env. Sci. Health, **A 13**, 1 and 7 (1978)

3 G. Martin (ed.), *Point sur l'Epuration et le Traitment des Effluents.* (Technique Documentation Lavoisier) 1982 p. 67

4 US Patent 4 543 190 (1985), Modar Incorporation; 14 Tech. Circle Natick, MA 01760 U.S.A.

Water Management in Power Stations

L. Pelloni, A. Kyas and I. Reimer
Brown Boveri, Baden, Switzerland

Abstract

Society realizes and recognizes the decrease in quality and quantity of water resources as a growing problem. This trend forces the utilities – by far the largest consumers of water – to manage the water streams in a power plant in a way that is economical and safe for the environment.

For existing power plants, there is increasing incentive to use less water, for example by recycling ("zero discharge"). As a first step, we give a full inventory of mass balances in the diverse water subsystems. The aim is to convert the numerous waste waters into clean water and compact products for easy disposal.

As an example, a systematic recording of the water streams in a power plant is shown. The areas for potential improvement are identified and possible corrective measures are presented.

Each year, about 100,000 tons of chlorine are applied world-wide to prevent biofouling in the condenser cooling water system. Chlorine forms non-degradable and toxic chlorinated organic compounds. This has led to regulatory restrictions on chlorine emissions. The application of ozone as a substitute for chlorination is a viable solution to this problem. Results of ozonation experiments using a pilot plant in a system with a cooling tower are presented. The effect of ozone on chemical and biological parameters and the corrosion characteristics of various materials were studied on-site.

1. Introduction

Society recognizes that the decrease in the quality and quantity of water resources is a growing problem. This means that we can no longer take an abundant supply of good water for granted. We estimate the world-wide consumption of water by utilities – excluding hydroelectric power – to amount to about 1,000 million cubic meters per day. This corresponds to the lifetime consumption of 100,000 inhabitants in an industrialized region. Thus, being among the largest consumers of the precious raw material water, power plants aim to manage their water system in a way that is economical and safe for the environment.

Fig. 1 depicts a power plant system and its principal inputs and outputs. The vertical streams represent input/output of mass and energy. The horizontal arrows indicate exchanges of information, showing the economical and legal boundary conditions. The various processes represent the key elements of the system and are situated at the intersections between the different mass streams. Optimization of water and chemicals consumption is therefore treated as an integrated problem. This means studying the overall effect of the individual processes on the system and on the environment.

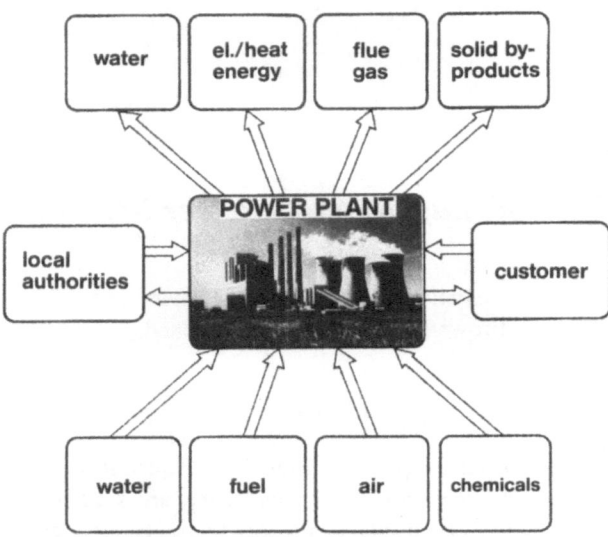

Fig. 1:   Power plant system with main input and output flows:
          • vertical direction: mass and energy
          • horizontal direction: information

Table 1:   Water Management: Potential savings in a 750 MWe coal fired power
          station.

Water Saving Technique	Limiting Factor	Potential Saving [10^6m^3/year]
Re-use and Recycle		
• Cooling tower blowdown for flue gas desulfurization	thresholding effect corrosion	0.5 - 1
• Bottom ash dewatering for coal dust suppression		1 - 1.5
• FGD waste water for fly ash handling		0.2 - 0.4
Treatment for disposal		
• Low volume wastes: Extraction of toxic inorganics (heavy metals)	thresholding effectless or no effect	disposal charge for 0.1- 0.3 million m³/year

2. Problem

Bearing in mind the philosophy "think about the end at the beginning", the major tasks achieving the goals of economical and environmentally safe water management may be summarized as follows:

- Supply the individual water-consuming processes with water of a quality which is only as good as required by the process[1]. This could mean either direct re-use or recycling with intermediate treatment, rather than using fresh water. Table 1 gives a general example for a 750 MWe coal fired power plant.

- Review and select water treatment chemicals. Consider the overall economical aspect. For example, the possible increase in discharge costs as a result of producing harmful secondary by-products or hazardous wastes.

In general, today's power plants embody process technology commercialized 5-10 years ago. In the meantime, consumption of water has considerably increased world-wide and tighter waste water regulations have been issued. Thus, economic parameters have changed and processes not considered feasible in the last decade, have become attractive today (e.g. membrane processes[2]). Moreover, high availability of process control hardware on the market now and improved mathematical models for treatment steps allow easier and better supervision and therefore help to reduce operating costs.

3. Solutions and Practical Implementation

The following two examples demonstrate that even today, with different economical and ecological standards, water demand and water processing costs in modern power plants can be cut and effluent quality can be improved.

3.1. Economical Management of Water Streams

3.1.1. Goal

As pointed out in the introduction, reducing waste water release means recycling the water using, if necessary, clean-up steps. This procedure addresses primarily the problem of quantity by reducing the amount of water required. Fig. 2 illustrates this philosophy.

The aim is to convert the numerous waste waters into water of re-usable quality. The resulting sludges and by-products must leave the system in a form allowing easy disposal[3].

3.1.2. Method

In order to find the areas where improvements are possible, an inventory of mass balances has to be given in the various water subsystems. The procedure shown in Fig. 3 is based on practical experience and is structured as follows:

- Define the water subsystems   } (actual situation)
- Classify the waste water streams }
- Establish general guidelines   } (target)
  in the form of a checklist   }

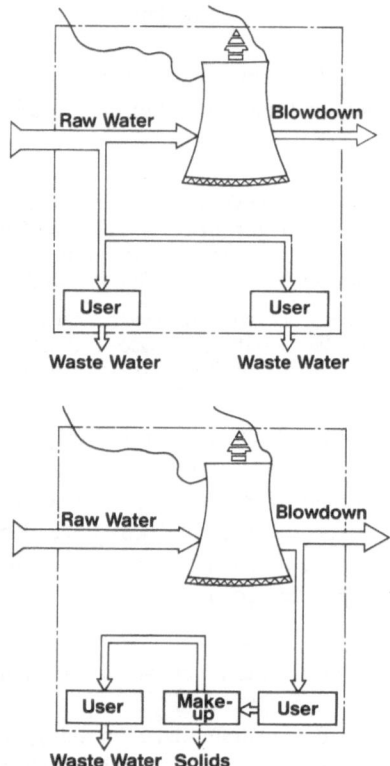

Fig. 2:   Top: Conventional distribution of raw water
           Bottom: Arrangement with integral water management.

The water subsystems are assigned to the following plant areas:

1) Turbine house
2) Boiler house
3) General area of power station
4) Rain run-off
5) Sanitary

For each subsystem, basic information, flow rates and quality of waste water are summarized in a data sheet.

For the classification: 20 different waste water types are defined (Type A to T). Table 2 shows a sample from a data sheet and from the classification criteria respectively. The flow and quality of the waste water depend on the operating conditions of the plant[3]. Further, a distinction is made between batch and continuous waste water discharge.

Guidelines: in dealing with waste water, better efficiency is achieved if the streams are chemically similar. The guidelines recommend joining waste water flows of the same type (quality), and storing, re-using and/or re-processing them[4].

As an example, the guidelines for a boiler house subsystem are shown in Fig. 4.

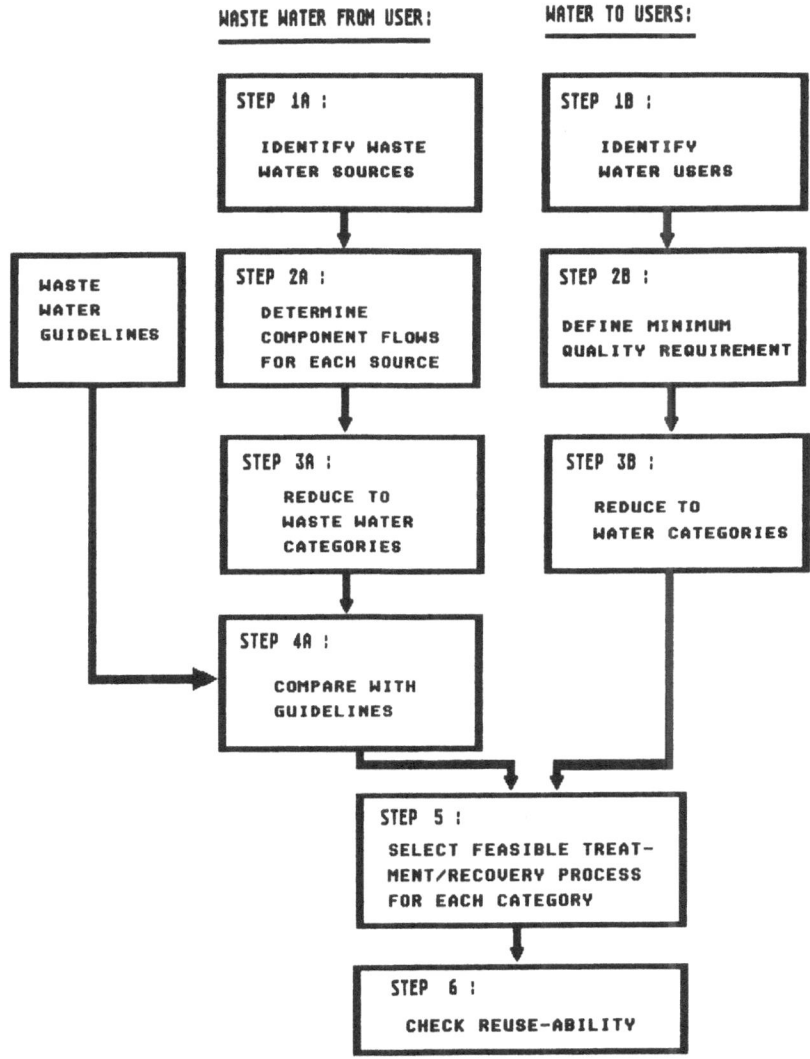

WASTE WATER FROM USER: WATER TO USERS:

STEP 1A :
IDENTIFY WASTE
WATER SOURCES

STEP 1B :
IDENTIFY
WATER USERS

WASTE
WATER
GUIDELINES

STEP 2A :
DETERMINE
COMPONENT FLOWS
FOR EACH SOURCE

STEP 2B :
DEFINE MINIMUM
QUALITY REQUIREMENT

STEP 3A :
REDUCE TO
WASTE WATER
CATEGORIES

STEP 3B :
REDUCE TO
WATER CATEGORIES

STEP 4A :
COMPARE WITH
GUIDELINES

STEP 5 :
SELECT FEASIBLE TREAT-
MENT/RECOVERY PROCESS
FOR EACH CATEGORY

STEP 6 :
CHECK REUSE-ABILITY

Fig. 3:   General procedure for improvement of water system management.

3.1.3. Practical Example

In an initial phase, the waste water system of a new coal-fired power plant (300 MWe) was audited according to steps 1A to 4A of the procedure described in Fig. 3. The configuration of the waste water systems of the power plant is shown in Fig. 5. In total, 72 single waste water sources were identified in the five areas. The different streams were classified into 14 types. For a considerable number of sources, neither the composition nor the flow-rate was known. The missing information was acquired by either on-site measurements or by estimating from experience and comparing with data from the literature.

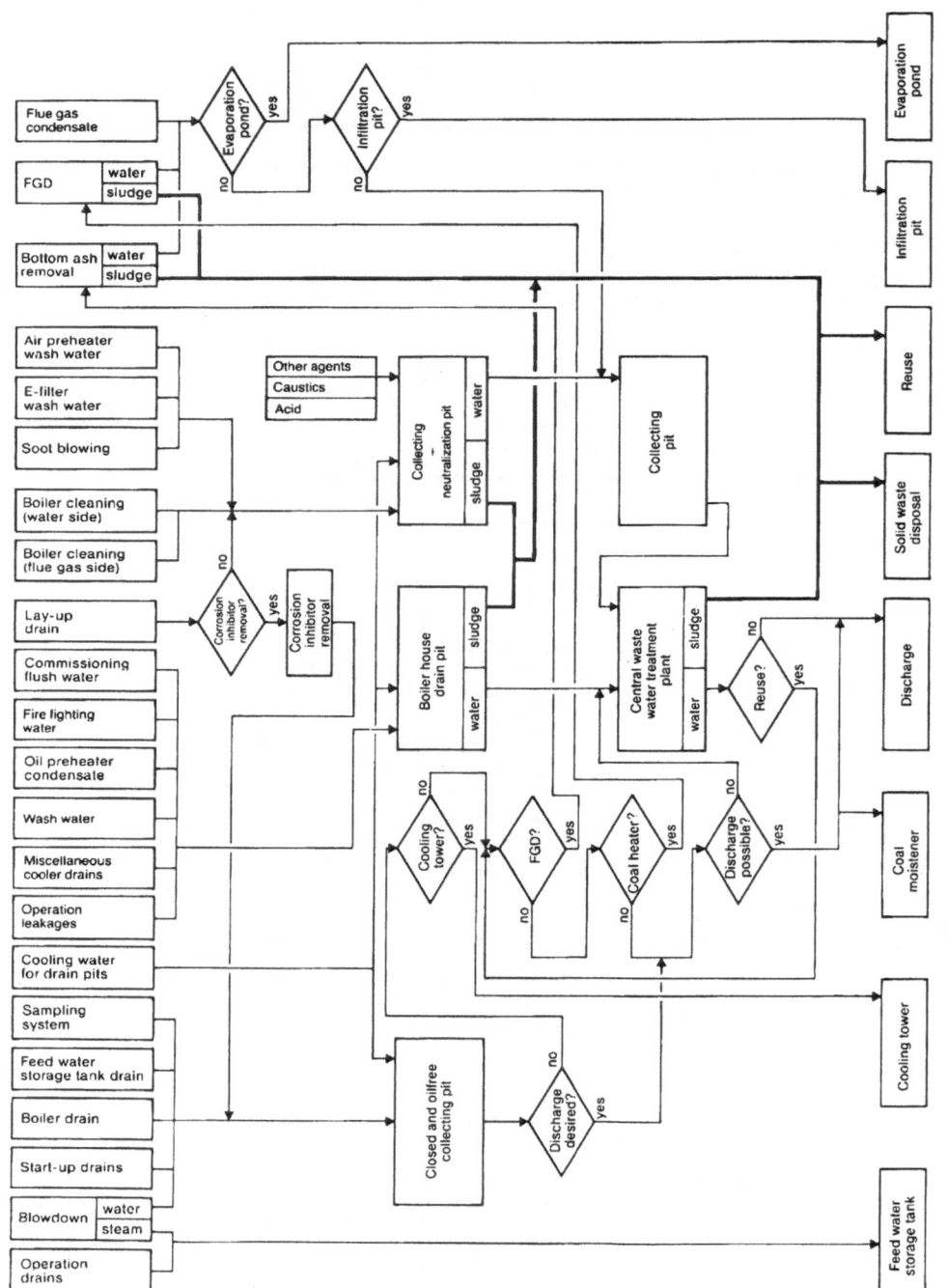

Fig. 4:   Guidelines for boiler house waste water system.

Table 2:　Top:　　　Selection from data sheet　.
　　　　　 Bottom:　Selection from classification criteria list.

Wastewater type	Flow rate m³/h		Water quality				Plant status				Fre-quen-cy		Remarks
Legend : NO = norm.operation RD = Rundown N = Frequency y = year m = month d = day	aver.	max.	°C	Type	Oil	Solids	Normal	Startup	Shutdown	Emergency	Permanent	Interned.	
1. Operation Drains	0.06 x MW		100	B			x				x		
2. Continuous Blowdown			1)										
Startup (Drum- and once-Through-boilers)	0.2 x MW	0.3 x MW	100	B D		x							
Norm.Operation (Drum boilers only)	0.02 x MW	0.05 x MW	1) 100	B D									

Parameter	Unit	Wastewater type				
		A	B	C	D	E
pH		6.5 - 7.5	8 - 10	6 - 10	8 - 11	6.5 - 7.5
Conductiv.	µS/cm	<2	3 - 20	10 - 200	5 - 100	400-5000
Susp.Sol.	mg/kg		<1	~ 300	1 - 10	~ 1
TDS	mg/kg			~ 100	~ 50	2500
Oil	yes/no	no	no	yes	no	no
TOC	mg/kg					<10
COD	mg/kg					<50
Tot.Hardn.	°dH					3°
Ammonia	mg/kg		0.2-10	0.2-10	0.2-10	
Hydrazine	mg/kg		<0.1	<0.1	<0.1	
Chloride	mg/kg					
Fluoride	mg/kg					

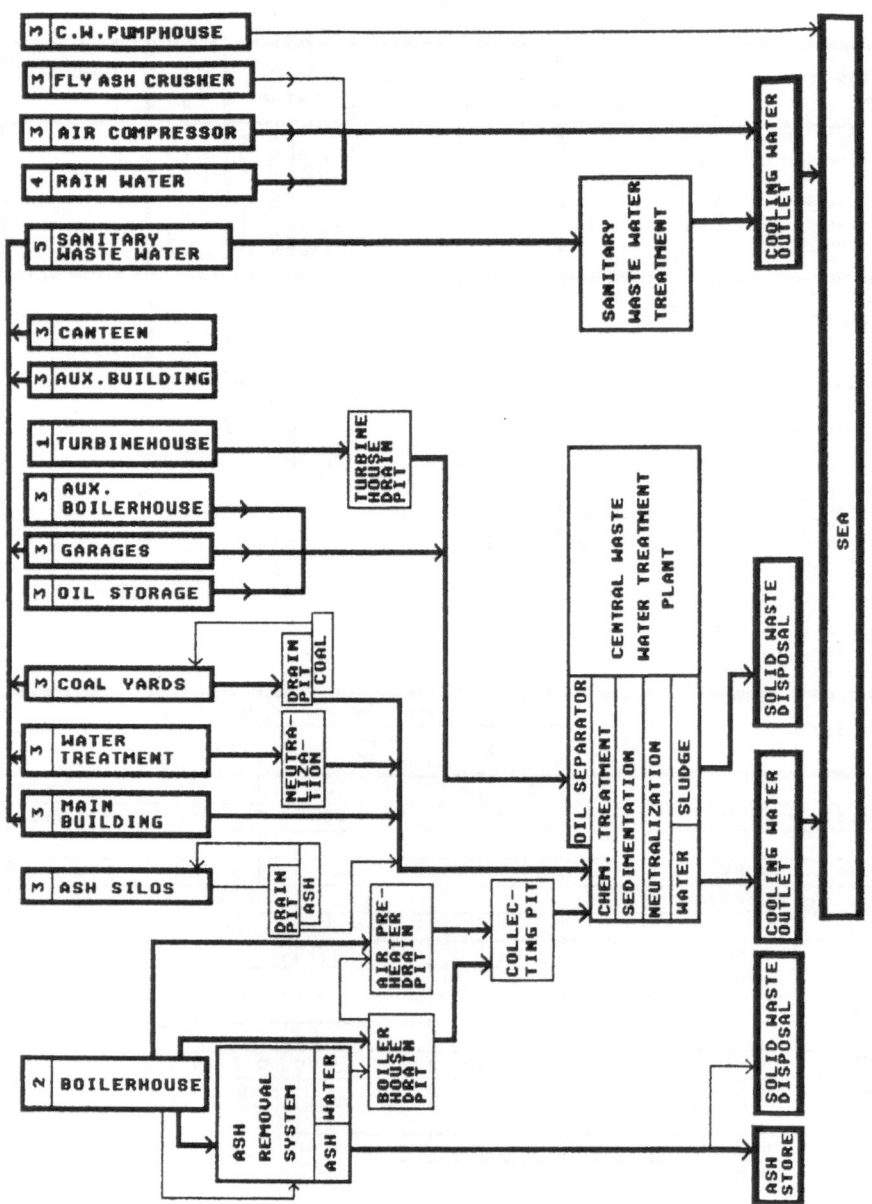

Fig. 5:   Principal waste water sources of a 300 MWe coal fired power plant.   Subsystems: 1 = turbine house, 2 = boiler house 3 = general area, 4 = rain run-off, 5 = sanitary

The data were classified according to the guidelines shown in Fig. 4. Potential areas for improvement were found. A representative selection of these results is shown in Fig. 6 for the boiler house waste water system.

The following are typical examples of inefficiency which were identified in this subsystem:

1) Uncontrolled quantities of mains water are added to the start-up-tank, in order to cool its content.
    Improvement: install cooling coils in the start-up-tank.

2) Oil from floor drains contaminated water collected in the air-preheater holding pond. This led to inefficient precipitation/flocculation in the central waste water treatment plant.
    Correction: install oil separators in drains[5] or connect outlet of holding pond to central oil separator.

3) During start-ups, too much waste water from the boiler house holding pond flowed into the central treatment plant because more mains water was added. This led to a temporary collapse of the waste water system.
    Correction: because the water in the holding-pond has a relatively high quality at normal operating conditions, disconnect it from the waste water treatment plant and discharge it directly or re-use it. Install equipment for removal of corrosion inhibitor at blow-down from boiler and from closed cooling-water cycle.

3.1.4. Conclusion

The results show that improvements are possible even in a new power station. Even the first steps, namely identifying the waste water streams and their characteristics, led to preventive measures being taken. These improved availability of the plant as well as fulfilling the discharge regulations.

3.2. Environmentally Safe Cooling Water Treatment

3.2.1. Goal

The largest water-flow in a power station is that through the condenser. Here, improved ecological compatibility (i.e. *quality*) rather than reducing quantity, is the crucial point. Regardless of the type of cooling system employed (open, closed, once-through), the water must be treated in order to prevent deposits and corrosion in the piping system. Deposits arise as a result of:

*   precipitation and crystallization of salts of low solubility (scaling);
*   sedimentation and adsorption of suspended solids on surfaces;
*   higher organisms entering the system and growing there.

However, since we are concerned with various mechanisms operating simultaneously, it is generally not sufficient to apply a single remedy, but rather a combination of different measures. We will consider as an example, only biofouling control by use of biocides.

Table 3 lists the methods known and applied today. Chlorination is the most common type of disinfection of cooling water. In the USA, about 10,000 to 100,000 tons of chlorine are used each year for that purpose[6]. The points of chlorine injection are shown in Fig. 7.

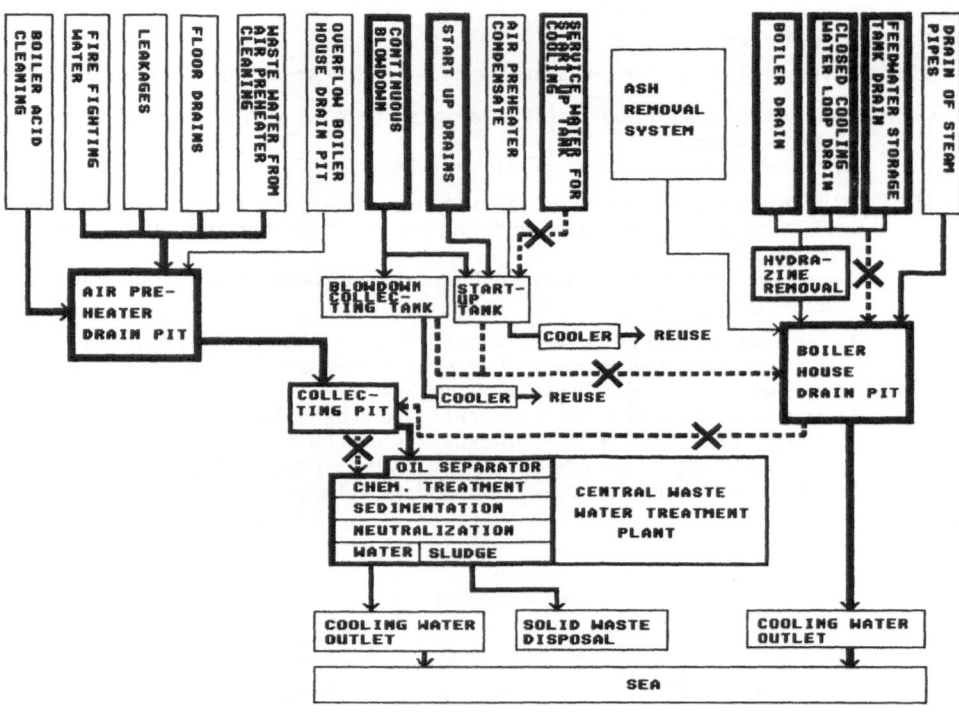

Fig. 6:   Boiler house waste water system and points of improvement.

Table 3:   Traditional methods to control biofouling.

Dosage of biocide	• Chlorine/hypochlorite • Hydrogen peroxide • UV • Various trade products	2 - 5 mg/l 10 - 120 mg/l
Surface as biocide	• Copper alloys	
Thermal method	• Recirculation of cooling water without back-cooling	T = 30 - 50 °C (monthly)
Physical method	• Screens • Sponge rubber balls • Shovel out	

Various references in the literature[2] are made to the toxicity of chlorine for aquatic organisms and its tendency to form chloro-organic compounds. This has led many authorities to limit the chlorine emission or to specify rapidly degradable biocides.

Ozone can be considered as a potential alternative to treatment with chlorine. Its use has a number of advantages:

*   ozone is a strong oxidizing agent;
*   it is produced on site so that there are no transport or storage problems;
*   unlike chlorine, it does not form any environmentally hazardous chloro-organic compounds;
*   ozone degrades rapidly, no removal of residual biocide is necessary.

In the last 10 years, reports have appeared in the literature concerning the use of ozone as a biocide. This information, particularly with respect to level of ozone dose, points of injection and effectiveness, is sometimes contradictory. Therefore we decided to perform our own tests in order to gain experience in this area. Seawater-cooled systems were not considered since, just as with chlorine, ozone forms hypo-bromite and bromate by reaction with bromide. Ecologically, both oxidation products are unacceptable[8,9].

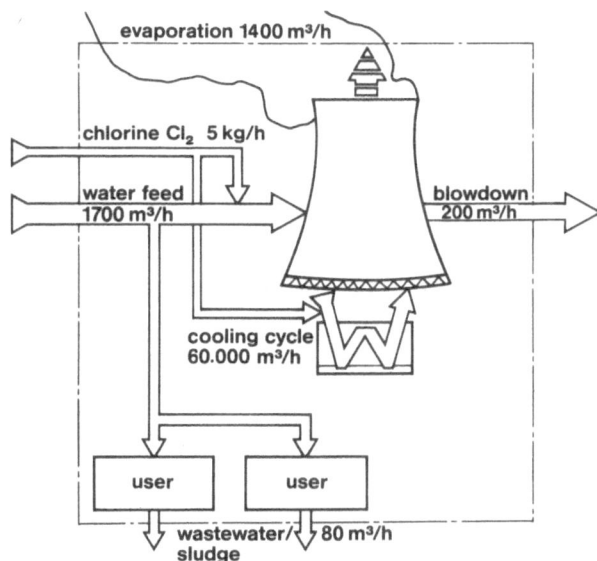

Fig. 7:   Chlorine injection and typical water flow rates for a 750 MWe coal-fired power plant.

3.2.2. Experimental

Cooling-water system (Fig. 8). In one of BBC's electronics factories, cooling-water is pumped to various users from two interconnected cooling-water tanks at the rate of $10 - 12 \text{ m}^3/\text{h}$. The water is passed through an induced draft cooling tower. Water losses are replaced with softened make-up water. Before ozone was employed as biocide, a commercially available antifouling/antiscaling solution was added to the cooling water. The main components and all pipes of the cooling-water system are made from stainless steel or polyethylene. However, some heat exchangers have copper alloy tubes.

Ozone contacting system. A sidestream of about 10% of the cooling-water flow is taken from the cooling-water tank. Ozone is introduced into this flow through a nozzle and thoroughly mixed with the water in a static mixer. The ozone-containing water is then reintroduced at the base of the tanks. A maximum concentration level of 0.8 g/m^3 of ozone is added to the cooling water bypass, corresponding to an ozon dose of $\sim 0.05 \text{ g/m}^3$ with respect to the total circulating cooling water flow.

A Brown Boveri ozone generator was used to supply the ozone. The concentration of ozone in the feed gas was monitored photometrically and the aqueous ozone concentration was measured by iodometry.

Algae growth test. In order to observe the effectiveness of our biocide treatment, two glass tubes (diameter 3 cm, length 70 cm) containing glass Raschig rings were installed. In one tube, green algae were introduced into standing water. This water is replaced weekly with fresh make-up water. Water from the cooling-water tank is circulated through the other tube. Both columns are illuminated with artificial sunlight for 12 hours per day.

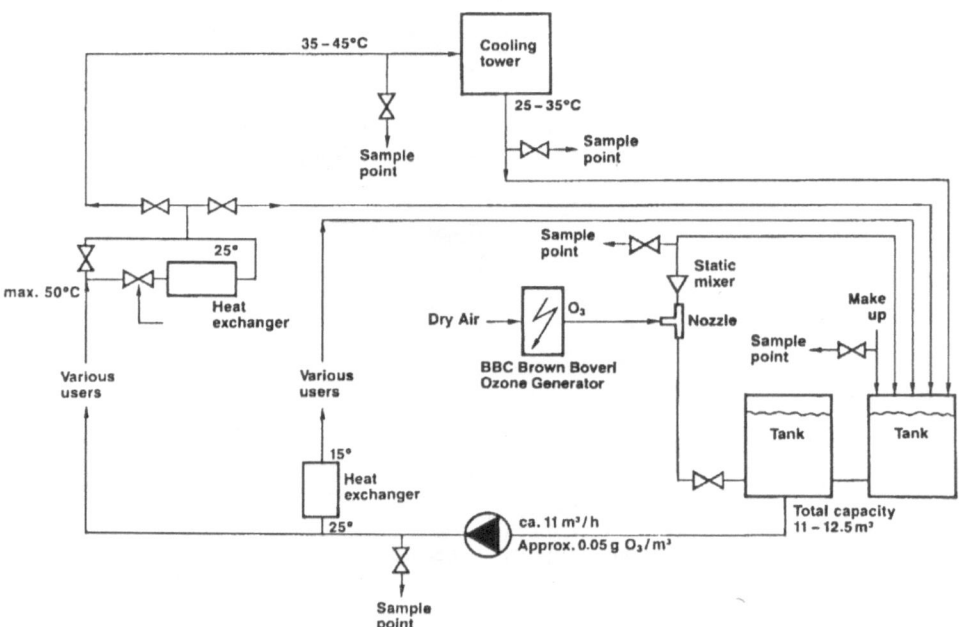

Fig. 8:   Flow sheet of cooling water system.

3.2.3. Tests Performed

In order to perform chemical, physical and microbiological investigations, samples were taken, as required, from the following places:

- make-up water (softened fresh water)
- at discharge of the circulating pumps (cooling water to the users)
- inlet to the cooling tower
- outlet from the cooling tower
- sidestream after the static mixer

The chemical and physical tests were performed according to recognized standard analytical methods. Millipore testers were used for the microbiological investigations. The determination of algae in the cooling water was performed on an enriched sample under the microscope. Samples (coupons) of various metals used as standard construction materials in cooling systems were exposed to the ozonized water of the sidestream as well as to the cooling-water stream at the tank outlet to check their stability and corrosion behaviour.

Table 4: Water quality parameters before and during treatment with ozone (A: before ozone addition, B: after 3 days ozonozation, C: After 6 months ozonozation).

Test Period Water type	A Make-up	B Make-up	A Cooling	B Cooling	C Cooling
pH-value	7.68	7.77	8.82	8.77	8.29
Conductivity [μS/cm]	447	456	680	688	515
$KMnO_4$ value [mg/l]	26.6	16.7	66.5	39	3.8
TDS [mg/l]	320	303	684	603	341
Turbidity [NTU]	-	0.09	0.23	0.20	-
Ammonia [mg/l]	0.028	0.011	0.093	0.129	0.032
Total hardness [mg $CaCO_3$/l]	10.1	10.3	23.6	19.5	3.6
Calcium [mg/l]	2.8	2.91	7.0	5.84	0.96
Magnesium [mg/l]	0.76	0.74	1.49	1.20	0.29
Copper [mg/l]	<0.01	<0.01	0.51	0.47	-
Iron [mg/l]	<0.001	<0.001	0.02	0.04	0.05
Chloride [mg/l]	14.4	14.6	94.5	79.8	17.8
Nitrate [mg/l]	-	-	-	-	14.4
Sulfate [mg/l]	22	22	27	23	29
Silica [mg/l]	6.9	5.8	3.4	3.4	5.9
Total count [ml^{-1}]					
24 h	100	-	0	0	0
48 h	-	-	-	-	1

Table 5:   Costs of ozone vs. chlorine for biofouling control.
    Base: • 600 MWe fossil fueled power plant
                   • Recirculating water for cooling tower 55′000 m³/h
                   • Hypochloride as solution containing 13 wt% chlorine

	Chlorine	Hypochlorite	Ozone
Consumption [kg/day]	190	1,500	65
Investment costs [SFr.]	400,000	100,000	480,000
Operating costs [SFr./360 days]	120,000	260,000	100,000

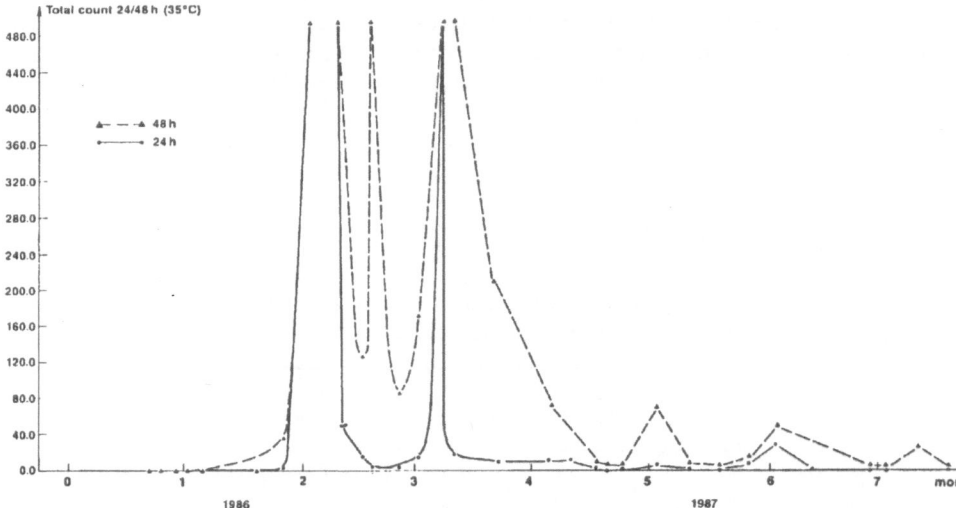

Fig. 9:   Total count in the cooling water at tank outlet.

3.2.4. Results and Discussion

Table 4 shows the results of water analysis. Samples were taken from different sampling points and at various times during the test phase.

Cooling water quality: the following significant changes were observed:

Total count (TC): the number of colonies was zero counts/ml up to the end of the 2nd month but then increased to an uncountable level (Fig. 9). The cause was an interruption of the water circulation, which caused an influx of raw water. A temporary increase in the ozone dose reduced the TC values. Reduction of the ozone dose resulted in a renewed increase in the TC. The ozone addition was increased once more in order to reduce the TC values. Now the TC varied typically between 2 and 50 counts/ml respectively at an ozone dosage of max. 0.8 g/m³ in the sidestream and 0.05 g/m³ in the circulating water.

Conductivity: at the beginning, a decrease in the conductivity occurred (Fig. 10) due to

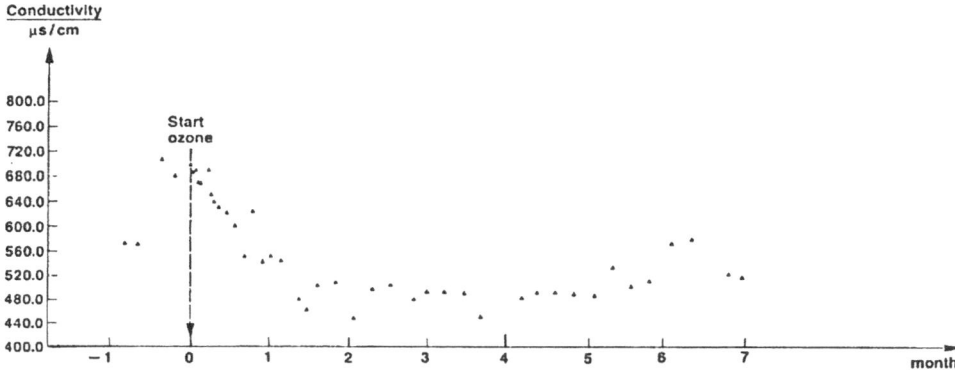

Fig.10: Conductivity of cooling water at tank outlet.

decreasing evaporation and slightly higher blow-down. Based on earlier experience with the present cooling circuit, an increase in the conductivity to about 1700 μS/cm can be used as an indicator for biofouling in the heat-exchanger tubes.

Oxidability (KMnO$_4$-values): a decrease of more than 90% was measured after 6 months of ozonation. The reasons for this decline were: no dosage of commercial antifouling solution; slightly higher blow-down and the oxidative effect of ozone.

Ozone: was not detectable in the circulating water.

Biological growth: there was no sign of slime accumulation/growth either in the tank or in the internal parts of the cooling tower. Further no growth could be detected in the glass column with circulating water, whereas green algae bred in the second control column.

Corrosion: no significant change in corrosion behaviour before and after ozonation has been observed so far for the materials in contact with the treated water.

Ozone emission: no ozone smell was detectable in the exhaust air of the cooling tower (threshold value ~ 20 ppb).

3.2.5. Conclusion

On the basis of more than 8 months of operating experiences, it can be concluded that ozone acts successfully as a biocide. A general cost comparison based on this initial operating experience, is presented in Table 5, and extrapolated for a 600 MWe power station.

Our results show that ozone treatment is an economical alternative to chlorine addition. Furthermore, environmental and safety aspects favour the use of ozone.

4. Final Remarks

These two examples demonstrate that the ecological balance for power plants can still be improved. In many cases, such measures also lead to increased overall efficiency and availability.The development of the methods and of the technology to improve water management in power plants requires profound know-how of both water technology and power generation.

References

1   Martin, H.J. and Miller, G.R., A zero discharge steam electric power generating station. Journal AWWA **5** (1986) 52-58.
2   Dobias, J., Schütze, R. and Stahl, M., Betriebswasser-Aufbereitung unter Anwendung der Umkehrosmosetechnik. VGB Kraftwerkstechnik **65**, 10 (1985) 959-964.
3   Strauss, S. D., Waste water management. Power, June (1986) 1-16.
4   Goldman, E. and Kelleher, P. J., Water reuse in fossil-fueled power stations. The Electric Power Industry (1973) 240-249.
5   Dallmann, C.H., Central wastewater treatment for a coal fired central station. AICHE Symposium Series **76**, 197 (1980) 215-224.
6   Malés, R., Emissions from power plant cooling system. EPRI Journal, March (1986) 45-46.
7   Wunderlich, M., Oekologischer Aspekt der Kühlwasserbehandlung. Technische Mitteilungen **78**, 9 (1985) 448-451.
8   Crecelius, E.A., Measurements of oxidants in ozonized seawater and some biological reactions. J. Fish. Res. Board Can. **36**, 8 (1979) 1006-1008.
9   Stewart, M.F., Blogoslawski, W.J., Hsu, R.Y., Helz, G.R., By-products of oxidative biocides: toxicity to oyster-larvae. Marine Pollution Bulletin **10**, 6 (1979) 166-169.

Discussion

Chairman: W.A. McRae
Zürich, Switzerland

M. Campagna   •   What would you have to add to your system in order to get rid of the side-products of chlorination, i.e. chlorinated hydrocarbons?

L. Pelloni   •   Of course there are technical solutions to the problem of removing chlorinated hydrocarbons. I'm not familiar with any feasible process for the big flow rates of water involved (1000 - 10,000 m^3/h). Any possible process for removing the chlorinated hydrocarbons, such as absorption or stripping, will lead to an increase of treatment costs of about two orders of magnitude.

H.P. Klein   •   Stripping just produces a phase change, because then you have the hydrocarbons in the air. The only solution is not to use chlorine.

L. Pelloni   •   I agree.

J.A. Redondo   •   I would like to know if you implemented the concept of integral water management in a power plant already, and in what have the savings been?

L. Pelloni   •   The investigations I've presented here were done in a coal fired power plant. The savings were about twice the costs of the investigations. To give you an idea: the necessary investigations will take about one to two months with two people and perhaps a local laboratory. The measures taken will start to show a return in one to three years. It's a small cost difference and the problem is to convince the utilities that there is a saving over a longer time period. As an example: if, for a 750 MWe coal-fired power plant, the cooling tower blow-down is used for a flue-gas desulfuration instead of being rejected, the saving is 10^6 m^3 times the price of the water. This means perhaps $ 200,000; the investigation costs about the same or less. So, in that case, after one year the expenses will be covered. We are convinced that the procedure is quite cost-effective and of couse we are not the only ones doing it. In many arid areas with high water costs of 1 $/$m^3$ or more like in South Africa, this was already done five years ago.

Subject Index